Novel Ceramic Materials

Edited by R. Saravanan

The book presents a number of novel ceramic materials that have great potential for advanced technological applications, such as microwave devices, communication instruments and memory devices.

The materials covered include piezoelectric ceramics, zirconia ceramics, doped NiO ceramic nanostructures, BST ceramics (Barium-Strontium-Titanates), manganite ceramics, Ce-doped $LaMnO_3$ and Sb-doped NKN (Sodium-Potassium-Niobates), as well as materials with ferrite structures, and with multi-ferroic structures

The materials were characterized experimentally by means of XRD (X-ray diffraction), SEM (Scanning electron microscopy), EDX (Energy Dispersive X-ray analysis), UV-Visible Spectroscopy, and VSM (Vibrating sample magnetometer). The results are discussed in terms of the structural characteristics of the various crystal structures, their special surface morphology, and their optical and magnetic properties.

Of particular interest is the determination of the electron density distribution (on the basis of XRD data and computerized evaluations). These data elucidate the atomic/electronic structure of the materials and make us understand the specific characteristics of these novel ceramics.

Novel Ceramic Materials

Edited by

Dr. R. Saravanan, M.Sc., M.Phil., Ph.D.
Associate Professor & Head
Research Centre and PG Department of Physics
The Madura College (Autonomous)
Madurai - 625 011
India

Copyright © 2016 by the authors

Published by **Materials Research Forum LLC**
Millersville, PA 17551, USA

Published as part of the book series
Materials Research Foundations
Volume 2 (2016)
ISSN 2471-8890 (Print)
ISSN 2471-8904 (Online)

Print ISBN 978-1-945291-02-9
ePDF ISBN 978-1-945291-03-6

Distributed worldwide by

Materials Research Forum LLC
105 Springdale Lane
Millersville, PA 17551
USA
http://www.mrforum.com/

Manufactured in the United State of America
10 9 8 7 6 5 4 3 2 1

Table of Contents

Preface

Many ceramic materials find use in different applications, such as microwave devices, communication instruments, memory devices, etc. and their potential use depends on the properties of the material. These properties can be determined by different characterization approaches. This edition presents experimentally determined properties of some of the most promising novel ceramic materials.

The experimental characterization of ceramic materials such as, piezoelectric ceramics like NKN, NKNS, ($Na_{1-x}K_xNbO_3$, $Na_{1-x}K_xNb_{1-y}Sb_yO_3$), zirconia ceramics like $Zr_{1-x}Li_xO_2$, doped NiO ceramic nanostructures, ferrite structures like $MgFe_2O_4$, multi-ferroic structures like $La_{1-x}Zn_xFeO_3$, BST ceramics like $Ba_{1-x}Sr_xTiO_3$, $BaTi_{1-x}Zr_xO_3$, manganite ceramics like $La_{1-x}Cr_xO_3$, are presented in this volume. Also, Ce (Cerium) doped $LaMnO_3$, Sb doped NKN ($Na_{1-x}K_xNb_{0.95}Sb_{0.05}O_3$) have been studied and analyzed.

The techniques used for the experimental studies are XRD (X-ray diffraction), SEM (Scanning electron microscopy), EDX (Energy Dispersive X-ray analysis), UV-Vis (UV-Visible Spectroscopy), VSM (Vibrating sample magnetometer) etc. Using these techniques the authors characterize the various materials as to their structure, surface morphological micro structure, optical properties, magnetic properties etc.

Using experimental XRD data, the electron density distribution has been determined between those atoms that are of special interest with respect to their bonding characteristics. This is insofar important as it is the distribution of the electrons in a given ceramic system that determines its chemical and physical properties and thus its potential technological applications.

CHAPTER 1

Chemical Bonding and Charge Density Imaging in $Ba_{0.2}Sr_{0.8}TiO_3$ Ceramics by Iterative Entropy Maximization

J. Mangaiyarkkarasi[1,*], R. Saravanan[2]

[1]PG and Research Department of Physics, NMSSVN College, Nagamalai, Madurai-625 019, Tamil Nadu, India

[2]Research Centre and PG Department of Physics, The Madura College, Madurai-625 011, Tamil Nadu, India

*mangai.jp@gmail.com

Abstract

The ferroelectric ceramic material $Ba_{0.2}Sr_{0.8}TiO_3$ has been synthesized by the conventional high temperature solid state reaction technique at 1400 °C for 5h. The X-ray diffraction technique and scanning electron microscopy were adopted to analyze the crystal structure and surface morphology of the sample. Cell constant and unit cell volume are derived from the Rietveld refinement. Average grain size is calculated as 43nm. Electron density distributions and chemical bonding natures between the atoms in the lattice site of $BaTiO_3$ were analyzed using the maximum entropy method (MEM). Charge density images and mid bond density values revealed the enhanced ionic character between Ba and O ions. Surface morphology is observed with scanning electron microscopy. Elemental compositions are further confirmed with energy dispersive X-ray spectroscopy.

Keywords

Barium Strontium Titanate, X-Ray Diffraction, Rietveld Refinement, Maximum Entropy Method, Electron Density.

Contents

1.   INTRODUCTION

Barium titanate ($BaTiO_3$) has always been a ceramic material of high research interest due to its unique properties such as chemical stability, high permittivity, low dielectric loss and high tunability [1]. By appropriate doping, the properties of $BaTiO_3$ can be dramatically manipulated to a great extent [2]. The enhanced piezoelectric properties of Ca doped ceramics due to the effects of sintering temperatures and poling conditions are reported by Cai-Xia Li et al. [3]. Photoluminescence behavior of Mn doped $BaTiO_3$ was explained by Sahoo et al. [4]. Positive temperature coefficients of resistivity (PTCR) of Pb doped materials are reported by Zhi Cheng Li et al.[5]. In the electronic industry ferroelectric ceramics are used for manufacturing dynamic random access memory (DRAM) [6]. Isovalent substitution of Sr^{2+} in the place of Ba^{2+} has shown great promise in the applications including phased array antennas and tuning elements. In the fabrication of microstrip patch antennas ferroelectric ceramic materials with high dielectric constant is extensively used [7]. The improvement in pyroelectric properties was observed by Zhang Guang Zu et al. [8] which is very helpful in the production of infrared sensors and detectors. The transition temperature can also be effectively controlled by adjusting Sr content. Substituting Sr ions (ionic size 1.13Å) for Ba ions (ionic size 1.35Å) systematically modifies the structural and electrical properties due to the A-site cation size differences. Smaller Sr ion causes the reduction in the average radius of A-site and stabilizes the cubic structure [9]. Bulk properties, electronic structure and magnetic properties of perovskites structures are also theoretically proposed by some researchers [10, 11]. In the present work, structural properties, electron density distributions, morphology, elemental composition of $Ba_{0.2}Sr_{0.8}TiO_3$ have been extensively studied. Increasing doping concentrations of Sr in $BaTiO_3$ lattice enhances the dielectric constant and drastically reduces the dielectric loss which is very useful in tunable microwave device applications. Particularly $Ba_{0.2}Sr_{0.8}TiO_3$ in the form of nanowires are suitable for ultra high density capacitors with fast discharge [12]. Electron density distributions and the bonding natures between the atoms are important to understand the properties of the material on an atomic level. The work concentrates on the chemical bonding and charge density imaging for the chosen system. This can be successfully evaluated with the maximum entropy method (MEM) [13].

2. SYNTHESIS AND CHARGE DENSITY ANALYSIS

2.1 SAMPLE PREPARATION

The ceramic sample $Ba_{0.2}Sr_{0.8}TiO_3$ was prepared by employing the conventional high temperature solid state reaction technique. High purity analytical grade barium carbonate ($BaCO_3$, 99.997%, Alfa aeser), strontium carbonate ($SrCO_3$, 99.99%, Alfa aeser) and titanium dioxide (TiO_2, 99.99%, Alpha aeser) were mixed well using an agate mortar according to the stoichiometry. The mixed powders were mechanically activated by a high energy ball mill with the speed of 500 rpm for 5 h, then the activated powder was placed in an alumina crucible and sintered to 1400 °C for 5 h using tubular furnace. Finally the sample was taken out and ground well for other characterization studies.

2.2 DATA COLLECTION

Powder XRD data were obtained in the 2θ range of 10°-120° with the step size of 0.02 from National Institute for Interdisciplinary Science and Technology (NIIST), Trivandrum, India using an X'pert-pro (Philips, Netherlands) X-ray diffractometer with the monochromatic incident beam of Cu-Kα (1.54056Å) radiation. SEM images corresponding to various magnifications (×1500 ×5000 ×10000) and EDAX results were collected at SAIF (Sophisticated Analytical Instruments Facility), Cochin University, Cochin, India.

2.3 STRUCTURE REFINEMENT USING POWDER XRD DATA

Fig. 1 shows the powder XRD pattern of $Ba_{0.2}Sr_{0.8}TiO_3$ ceramics. The existence of well defined and sharp peaks indicate that this ceramic material has a high degree of crystallinity at long range. Phase purity and cubic perovskite crystal structure of the sample were confirmed using the Joint Committee on Powder Diffraction Standards (JCPDS) (PDF #34-0411).

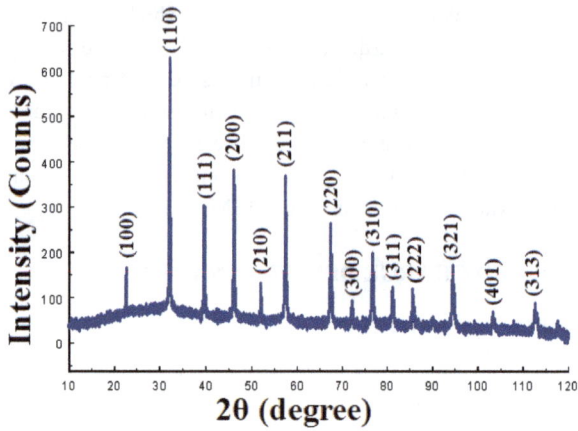

Fig. 1 The observed X-ray diffraction pattern of $Ba_{0.2}Sr_{0.8}TiO_3$

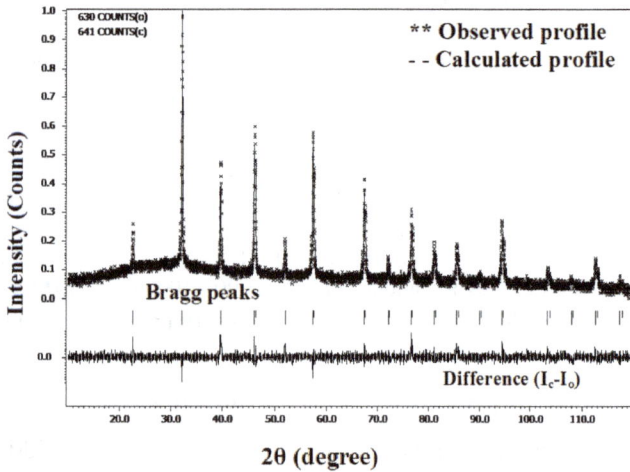

Fig. 2. Rietveld refinement plot of $Ba_{0.2}Sr_{0.8}TiO_3$

The raw XRD profile was subjected to Rietveld refinement using the software package JANA 2006 [14]. Structural parameters such as lattice parameters, atomic coordinates, composition of atoms, shift, background, scale factors and some other parameters were refined from the experimentally observed XRD profiles by comparing them with the theoretically constructed profiles.

The refinement was done with Pm-3m space group and position coordinates (0,0,0) for barium / strontium, (0.5, 0.5, 0.5) for titanium and (0.5, 0.5, 0) for oxygen [15]. Fig. 2 depicts the refined profile in which the dots represent the observed XRD profiles and continuous lines represent the calculated profiles. The difference between the observed profile and the calculated profile is shown at the bottom of the figure. The positions of the Bragg peaks are indicated by the small vertical lines below the fitted profile. Refined structural parameters and the reliability indices are given in Table 1.

Table .1. Structural parameters of $Ba_{0.2}Sr_{0.8}TiO_3$ from Rietveld refinement

Parameters	values
a (Å)	3.9270(14)
$\alpha = \beta = \gamma$	90
Cell Volume (Å^3)	60.56(23)
Density (gm/cc)	5.32(18)
R_{obs} (%)	5.12
wR_{obs} (%)	5.72
R_p (%)	8.99
wR_p (%)	12.11
GOF	0.90

In this table, the fitting parameters (R_p, wR_p, R_{obs} and GOF) indicate a good match between the refined and observed XRD patterns. Lattice constant and cell volume are found to be closely matched with the reported values [16].

2.4 GRAIN SIZE DETERMINATION

The average grain size of the synthesized ceramic material was calculated with full width at half maximum and Bragg angles obtained from XRD data by adopting the GRAIN software [17] using the Debye-Scherrer formula (t = 0.9λ / $\beta cos\theta$, where t is grain size, λ is wavelength of X-ray (λ=1.54056Å), β is the full width at half maximum and θ is the Bragg angle). The resultant average grain size is 43 nm which compares well with previous studies [18].

2.5 CHARGE DENSITY STUDIES

The analysis of the electronic structure of materials and the bonding between the atoms is an important part of materials characterization. Structure factors evolved from the Rietveld refinement were utilized for the estimation of charge density in the unit cell of $Ba_{0.2}Sr_{0.8}TiO_3$ through the maximum entropy method (MEM) [13]. Evaluation of charge density distribution is done using the software package PRIMA [19] and the 3-dimensional charge density distributions in the unit cell and 2-dimensional electron density maps are plotted along with the help of the visualization software VESTA [20]. Fig. 3 shows the 3-dimensional unit cell with electron density distributions by considering the same isosurface level of $1e/Å^3$. The positions of atoms and the electron density between the atoms are made clearly visible in this figure.

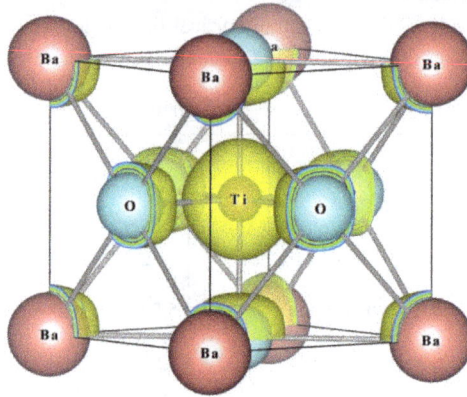

Fig. 3.3-dimensional electron density distribution of $Ba_{0.2}Sr_{0.8}TiO_3$

(a)   **(b)**   **(c)**

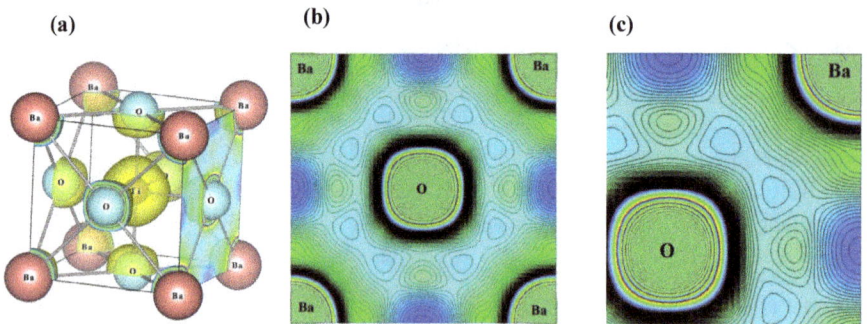

Fig. 4(a). 3D unit cell of $Ba_{0.2}Sr_{0.8}TiO_3$ with (100) plane shaded. (b) 2D electron density map on (100) plane (c) Enlarged view of Ba-O bond

The 3D unit cell of $Ba_{0.2}Sr_{0.8}TiO_3$ with the shaded (100) plane is shown in Fig. 4(a) and 4(b) represents the 2-dimensional electron density map on the (100) plane. Fig.4(c) represents the enlarged view of Ba-O bond on the (100) plane.

From the contour lines of charge density images it can be visualized that there is not much charge linkage between the Ba and the O atoms which is a sign of ionic nature between Ba and O because there is weak hybridization between the valance levels of barium and oxygen atoms [21].

(a)   (b)   (c)

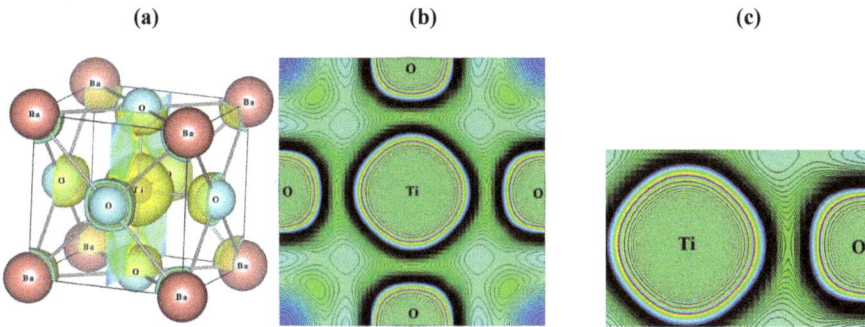

Fig. 5(a). 3D unit cell of $Ba_{0.2}Sr_{0.8}TiO_3$ with (200) plane shaded. (b) 2D electron density map on (200) plane (c) Enlarged view of Ti-O bond

(a)   (b)   (c)

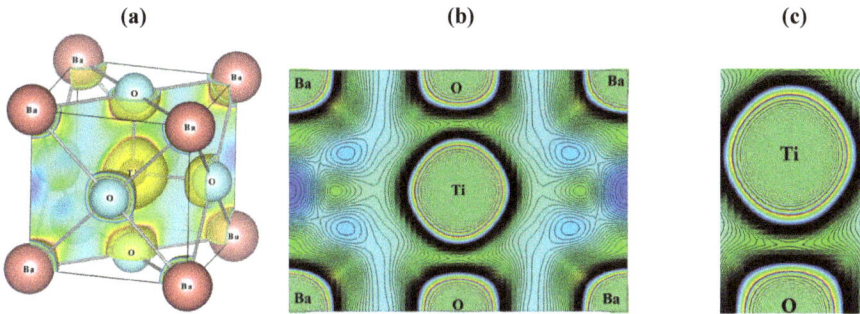

Fig. 6(a). 3D unit cell of $Ba_{0.2}Sr_{0.8}TiO_3$ with (101) plane shaded. (b) 2D electron density map on (101) plane (c) Enlarged view of Ti-O bond

Fig. 5(a) shows the 3D unit cell with (200) plane shaded and 5(b) represents the 2-dimensional electron density map. Fig.5(c) represents the enlarged view of Ti-O bond on the (200) plane. Fig. 6(a) shows the 3D unit cell with (101) plane shaded and 6(b) represents the 2-dimensional electron density map. Fig. 6(c) represents the enlarged view of Ti-O bond on the (101) plane. From these figures, explicit sharing of charges between Ti and O atoms are visible which suggests the covalent nature between the Ti and O. In order to analyze the nature of bonding charges involving between the atoms and to determine the bond lengths in detail, one dimensional electron density profiles are drawn.

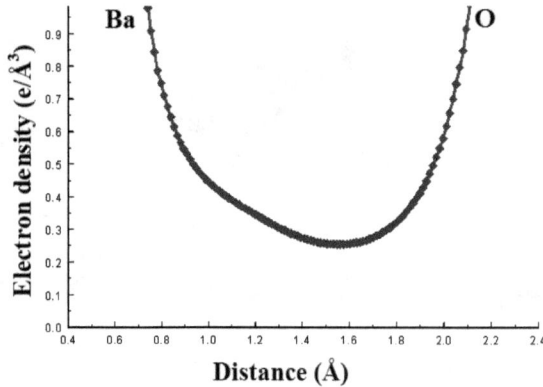

Fig.7. One dimensional electron density profile for (Ba-O) bond

Fig.7. represents the one dimensional electron density profiles corresponding to Ba-O bond and Fig. 8. shows the one dimensional electron density profile for Ti-O bond. Table 2 gives the bond lengths and mid bond density values between the bonds.

Table .2. Electron density values from 1D profiles.

Bond	Bond length (Å)	Mid bond density $(e/Å^3)$
(Ba-O)	2.776	0.279
(Ti-O)	1.963	0.639

8

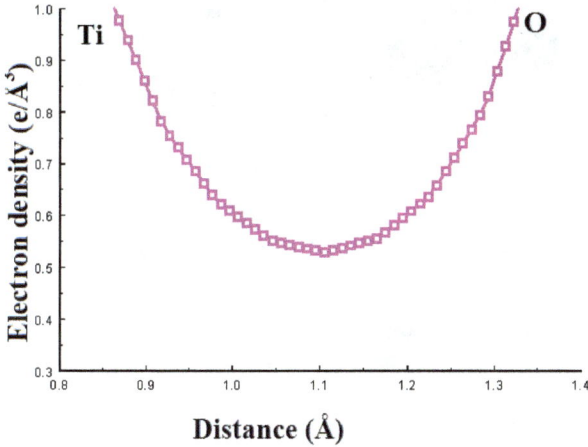

Fig. 8. One dimensional electron density profile for (Ti-O) bond

Mid bond density between Ba and O atoms is 0.279 e/Å^3, this confirms the ionic nature of Ba-O bond and 0.639 e/Å^3 between Ti and O atoms and endorses the covalent nature of Ti-O bond. The ionic character of Ba-O bond and covalent character of Ti-O bond is also reported by Gurgel et.al [22] through the density functional theory.

2.6   SEM/EDAX ANALYSIS

Surface morphology and the microstructure of the sample were investigated with scanning electron microscopy (SEM).

Table .3. Elemental compositions from EDAX

Element	Mass (%)	Atom (%)
Ba	52.68	33.89
Sr	43.24	43.59
O	4.08	22.52

Fig. 9. SEM image of $Ba_{0.2}Sr_{0.8}TiO_3$ at the magnification level of ×5000

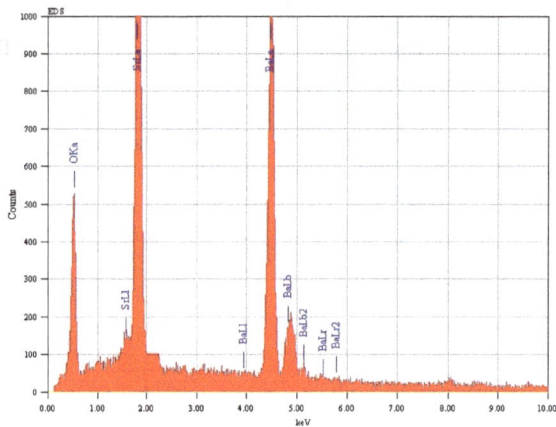

Fig. 10. EDAX spectrum of $Ba_{0.2}Sr_{0.8}TiO_3$

SEM image corresponding to the magnification×5000 is presented in fig. 9. which exhibits irregular microstructures with non uniform shapes. In order to decrease the surface free energy, atoms move from particles with smaller radii to those of higher radii, resulting in the formation of irregular grains [23] .The average particle size is estimated as 2 μm also matched with the earlier studies [24]. In order to investigate the chemical homogeneity of the sample an EDAX analysis was done. Fig .10. depicts the EDAX spectrum and table 3 exhibits the percentages of atoms present in the prepared sample thereby confirming the elemental compositions of $Ba_{0.2}Sr_{0.8}TiO_3$.

CONCLUSION

A single phased $Ba_{0.2}Sr_{0.8}TiO_3$ ceramic sample has been successfully prepared with the high temperature solid state reaction technique. The XRD results confirm the cubic perovskite nature of the prepared sample. From the Rietveld refinement cell parameter value is found to be 3.927 Å and the cell volume is 60.56($Å^3$). Observed structure factors are in close agreement with the structure factors derived from the Rietveld refinement. Average grain size is determined as 43nm. Dominating covalent nature of the Ti-O bond and ionic nature of the Ba-O bond are revealed from MEM charge density images and mid bond density values. Irregular shaped microstructures are observed in the SEM image. Chemical compositions of the sample are checked with the EDAX analysis.

REFERENCES

[1]   Shi Ai-hua, Yan Wen-bin, Li You-ji, Huang Ke-long, Preparation and characterization of nanometer-sized barium titanate powder by complex-precursor method, J. Cent. South Univ. Technol.15 (2008) 334−338.
       http://dx.doi.org/10.1007/s11771-008-0063-2

[2]   M.T. Buscaglia, V. Buscaglia, M. Viviani, P. Nanni, M. Hanuskova, Influence of foreign ions on the crystal structure of BaTiO3, Journal of the European Ceramic Society 20 (2000) 1997-2007.
       http://dx.doi.org/10.1016/S0955-2219(00)00076-5

[3]   Cai-Xia Li, BinYang, Shan-TaoZhang, RuiZhang, Wen-WuCao, Effects of sintering temperature and poling conditions on the electrical properties of Ba0.70Ca0.30TiO3 diphasic piezoelectric ceramics, Ceramics International 39 (2013) 2967−2973.
       http://dx.doi.org/10.1016/j.ceramint.2012.09.073

[4]   T. Sahoo, G.K. Pradhan, M.K. Rath, B. Pandey, H.C. Verma,S. Nandy, K.K. Chattopadhyay, S. Anand, Characterization and photoluminescence studies on hydrothermally synthesized Mn-doped barium titanate nano powders, Materials Letters 61 (2007) 4821−4823.
       http://dx.doi.org/10.1016/j.matlet.2007.03.053

[5]   Zhi Cheng Li, Bill Bergman, Effect of ageing on the electrical resistivities of (Ba0.69Pb0.31)TiO3 PTCR ceramic thermistors, Ceramics International 31 (2005) 375-378.

http://dx.doi.org/10.1016/j.ceramint.2004.06.021

[6] Fangyi Rao, Miyoung Kim, A. J. Freeman, Shaoping Tang and Mark Anthony, Structural and electronic properties of transition-metal/BaTiO3(001) interfaces, Physical Review B Volume 55, Number 20, 15 MAY 1997-II. *http://dx.doi.org/10.1103/PhysRevB.55.13953*

[7] Uma Shankar Modani, Gajanand Jagrawal, A survey on Application of Ferroelectric Materials for Fabrication of Microstrip Patch Antennas, International Journal of Recent Technology and Engineering (IJRTE) ISSN: 2277-3878, Volume-1, Issue-5, November 2012.

[8] Zhang Guangzu, Yi Jinqiao, Jiang Shenglin,Yu Yan, He Jungang, Liu Sisi, Zhu Dingyang, Zhang Ling, Role of internal stress on dielectric and dc bias field-induced pyroelectric properties of Ba0.68Sr0.32TiO3-(Ba0.68Sr0.32)2TiO4 for uncooled infrared detectors, http://www.paper.edu.cn.

[9] R. K. Zheng, J. Wang, X. G. Tang, Y. Wang, H. L. W. Chan, C. L. Choy, X. G. Li, Effects of Ca doping on the Curie temperature, structural, dielectric, and elastic properties of Ba0.4Sr0.6−xCaxTiO3(0≤x≤0.3) perovskites, Journal of Applied Physics 98, 084108 (2005). *http://dx.doi.org/10.1063/1.2112175*

[10] S. Piskunov, E. Heifets, R.I. Eglitis, G. Borstel, Bulk properties and electronic structure of SrTiO3, BaTiO3, PbTiO3 perovskites: an ab initio HF/DFT study, Computational Materials Science 29 (2004) 165–178. *http://dx.doi.org/10.1016/j.commatsci.2003.08.036*

[11] Hiroyuki Nakayama and Hiroshi Katayama-Yoshida, Theoretical Prediction of Magnetic Properties of Ba(Ti1−xMx)O3 (M=Sc,V,Cr,Mn,Fe,Co,Ni,Cu), Jpn. J. Appl. Phys. Vol. 40 (2001) pp. L 1355–L 1358.

[12] Hai Xiong Tang, Henry A.Sodono, Ultra high energy density nanocomposites capacitors with fast discharge using Ba0.2Sr 0.8TiO3 nanowires j. of Nanoletters 13 (2013)1373-1379.

[13] D.M. Collins. Electron density images from imperfect data by iterative entropy maximization. Nature., 298, (1982) 49-51. *http://dx.doi.org/10.1038/298049a0*

[14] V. Petricek, M. Dusek, L. Palatinus, The Crystallographic Computing System JANA 2006. (Institute of Physics, Academy of Sciences of the Czech Republic, Praha, 2000).

[15] R.W.G. Wyckoff, Crystal structures. 1 Inter-space publishers, London, 1963.

[16]   Noor Jawad Ridha, W. Mahmood Mat Yunus, S.A. Halim, Z.A. Talib, Firas K. Mohamad Al-Asfoor and Walter C. Primus, Effect of Sr Substitution on Structure and Thermal Diffusivity of Ba1-xSrxTiO3 Ceramic, American J. of Engineering and Applied Sciences 2 (4): (2009) 661-664.

[17]   R. Saravanan, Grain software (Personal communication) (2008).

[18]   Teresa Hungria, Miguel Alguero,Ana B. Hungria, and Alicia Castro, Dense, Fine-Grained Ba1-xSrxTiO3 Ceramics Prepared by the Combination of Mechanosynthesized Nanopowders and Spark Plasma Sintering, Chem. Mater., 17, (2005) 6205-6212.
        http://dx.doi.org/10.1021/cm0514488

[19]   A.D. Ruben, F. Izumi, Super-fast Program PRIMA for the Maximum-Entropy Method, Advanced Materials Laboratory (National institute for materials science, Tsukuba, Ibaraki, 2004).

[20]   K. Momma, F. Izumi, VESTA: a three-dimensional visualization system for electronic and structural analysis Journal of Applied Crystallography. 41, (2008), 653.
        http://dx.doi.org/10.1107/S0021889808012016

[21]   H. Salehi, N. Shahtahmasebi, and S.M. Hosseini, Band structure of tetragonal BaTiO3, Eur. Phys. J. B 32, 177–180 (2003).
        http://dx.doi.org/10.1140/epjb/e2003-00086-6

[22]   M.F.C. Gurgel, J.W.M. Espinosa, A.B. Campos, I.L.V. Rosa, M.R. Joya, A.G. Souza, M.A. Zaghete, P.S. Pizani, E.R. Leite, J.A. Varela, E. Longo Photoluminescence of crystalline and disordered BTO:Mn powder: Experimental and theoretical modeling Journal of Luminescence 126 (2007) 771–778.
        http://dx.doi.org/10.1016/j.jlumin.2006.11.011

[23]   M.Ganguly, S.K. Rout, T.P. Sinha, S.K. Sharma, H.Y. Park, C.W. Ahn, I.W. Ki, Characterization and Rietveld Refinement of A-site deficient Lanthanum doped Barium Titanate, Journal of Alloys and Compounds 579 (2013) 473–484.

        http://dx.doi.org/10.1016/j.jallcom.2013.06.104

[24]   S. Suasmoro, S. Pratapa, D. Hartanto, D. Setyoko, U.M. Dani, The characterization of mixed titanate Ba1-xSrxTiO3 phase formation from oxalate coprecipitated precursor, Journal of the European Ceramic Society 20 (2000) 309-314.
        http://dx.doi.org/10.1016/S0955-2219(99)00143-0

CHAPTER 2

Synthesis, Characterization and Charge Density Analysis of Lead Free Piezoceramics $Na_{1-x}K_xNbO_3$

S. Sasikumar, R. Saravanan

Research Centre and Post Graduate Department of Physics, The Madura College, Madurai-625 011, Tamil Nadu, India

Email: saragow@gmail.com; sasikuhan@gmail.com

Abstract

Solid solutions of lead free ceramics of $Na_{1-x}K_xNbO_3$ (x=0.01, 0.03 and 0.05) were prepared by the conventional solid state reaction method. The formation of perovskite structure of the prepared ceramics was confirmed by means of room temperature powder X-ray diffraction. The powder X-ray diffraction study confirmed that $Na_{1-x}K_xNbO_3$ (x=0.01, 0.03 and 0.05) adopted the orthorhombic structure. The crystal structures were refined using profile refinement. The electron density distributions of these crystals were analysed with the maximum entropy method (MEM) using powder X-ray diffraction data. Aggregated average particle sizes were evaluated using scanning electron microscopy. Elemental compositional analysis was carried out using energy dispersive spectrum. Optical properties are analyzed by UV-Visible spectrum.

Keywords

Crystal Structure, X-Ray Diffraction, Rietveld Analysis, Optical Properties, Polyhedral

Contents

1.   INTRODUCTION

Nowadays, more and more attention is being paid to the applications of potassium-sodium niobate (KNN) based ceramics after decades of development in basic material research [1-3]. KNN-based ceramics have been applied to lead-free piezoelectric actuators, sensors and transducers such as multilayer actuators, liner ultrasonic motor, speakers, touch sensor and transformer etc. [4-9]. (K,Na)NbO$_3$ (KNN) materials have been recently given considerable attention because of their good piezoelectric constant (d_{33}), a high Curie temperature (T_c), and environmental friendliness [10-13], they are considered as a promising candidate in the field of lead-free piezoelectric ceramics. However, it is difficult to obtain dense KNN ceramics with a high d_{33} value by using the conventional sintering method [11, 12]. Some attempts have been used to improve the density and piezoelectric properties of KNN-based ceramics, such as preparation method [13].

Alkaline niobate perovskites, in general, potassium sodium niobate (K,Na)NbO$_3$ (KNN) in particular has been widely studied. This is a potential candidate due to its high Curie temperature (420°C) and promising ferroelectric and piezoelectric properties comparable with PZTs [14]. In the present work, we have reported synthesis of Na$_{1-x}$K$_x$NbO$_3$ (x=0.01, 0.03 and 0.05) ceramics by the solid state reaction method, crystal structure, structural refinement using Rietveld refinement [15] and the electron density distributions analysed by the maximum entropy method (MEM) [16].

2.   SAMPLE PREPARATION

Na$_{1-x}$K$_x$NbO$_3$ (x=0.01, 0.03 and 0.05) ceramics were synthesized using solid state reaction method. Na$_2$CO$_3$ (99.99%), K$_2$CO$_3$ (99.99%), Nb$_2$O$_5$ (99.99 %) powders were used as starting materials. The powders weighted with their stoichiometric ratio. The powders were mixed by ball milling for 3 h with agate balls. The mixed powders were calcined at 900 °C for 4 h. The calcined samples were pressed into pellets of 15 mm diameter and thickness of 2 mm using hydraulic press. The pelletized samples were finally sintered at 1090 °C for 2 h in programmable furnace.

X-ray diffraction measurements at room temperature, used to investigate the purity of the perovskite phases were performed with a D8 Advance (Karlsruhe, Germany) diffractometer using CuKα radiation (λ=1.54056 Å), with step size of 0.02° and for 2θ=10-120°. The UV-Visible spectra give optical properties of the prepared sample using Varian, Cary 5000. Scanning electron microscope (JEOL JEM-2100F) coupled with EDS was used to analyse the microstructure and to check the chemical composition of the ceramic samples. The lattice parameters of the ceramics were refined by the Rietveld refinement [15] using the X-ray diffraction refinement program JANA 2006 [17]. The characterization measurements were analyzed at SAIF (Sophisticated Analytical Instrument facility), Cochin, Kerala, India.

3.   RESULTS AND DISCUSSION

3.1   POWDER X-RAY DIFFRACTION ANALYSIS

The XRD patterns of $Na_{1-x}K_xNbO_3$ ceramics with various x=0.01, 0.03 and 0.05 values sintered at 1040 °C are shown in Figure 1 (a). The XRD peaks have been indexed with orthorhombic crystal structure with space group Pbcm (JCPDS file no. 37-1493). All ceramics possess a pure perovskite structure without any other second phase indicating that the Na^+ ions were substituted by K^+ ions. Figure 1 (b) shows the enlarged (200) peak, which shows that the peak shifts towards the lower diffraction angles side of 2θ diffracting angle. It can be seen that the unit cell volume increased with the increase of K^+ content on $NaNbO_3$. It is expected that K^+ (1.33Å) should substitute for Na^+ (1.02Å) at A site [18]. The substitution on A-site may be leading to the expansion of the crystal unit cell. The increasing trend in the cell volume with the increase in K^+ concentration indicates the replacement of $NaNbO_3$ by potassium. The variations in the lattice parameters and cell volume have also been studied for different doping concentrations and are presented in table 1.

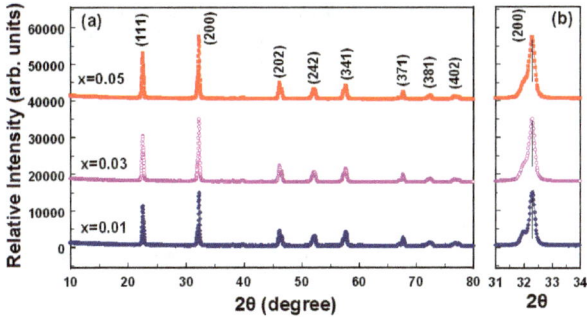

Figure 1 (a) X-diffraction patterns for $Na_{1-x}K_xNbO_3$ (b) enlarged (2 0 0) peak.

3.2 RIETVELD ANALYSIS

Figure 2 (a-c) depicts the observed and calculated profiles obtained by the Rietveld analysis [15] of the XRD data of $Na_{1-x}K_xNbO_3$ with x=0.01, 0.03 and 0.05 using the orthorhombic space group Pbcm. In the present work, the Rietveld refinement was performed through the JANA 2006 program [17]. In this technique, structural parameters, lattice parameters, peak shift, background profile shape and preferred orientation parameters are used to minimize the difference between a calculated profile and the observed data. The lattice parameters (a, c and V) and fitting parameters (R_p, R_{wp}) of the ceramics are shown in table 1. From table 1, small values of the reliability R_p, R_{wp} and GOF are obtained. This is confirmed by observing the difference pattern in measured and calculated XRD pattern. Table 1 shows small distortion in the lattice is mainly due to the substitution of Na^+ by K^+ ion. In table 1 has been observed that, there is significant increase in cell volume with K^+ doping ions and this is also attributed to the increase in the ionic radii.

Figure 3 shows the polyhedron representation of the $Na_{0.99}K_{0.01}NbO_3$ crystal structure generated with the JANA 2006 [17] program using the refined cell parameter, space group and atomic coordinate (xyz) of the atoms. The tetrahedral site, in which the Na atom is surrounded by four oxygen atoms in a regular tetrahedron and the octahedral site in which Nb atom is surrounded by six oxygen atoms are shown in figure 3. The iso-surface levels are suppressed for a better view of the tetrahedral and octahedral arrangement.

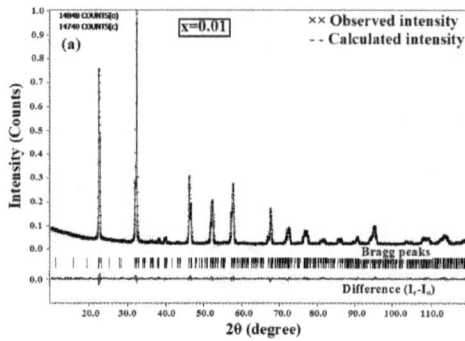

Figure 2 (a) Rietveld refinement profile for $Na_{0.99}K_{0.01}NbO_3$.

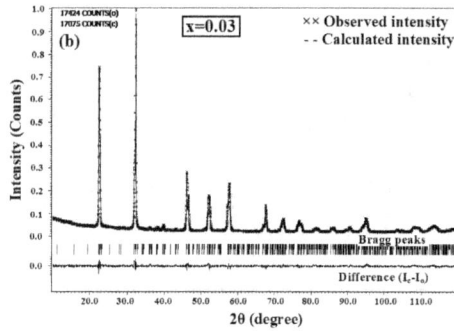

Figure 2 (b) Rietveld refinement profile for $Na_{0.97}K_{0.03}NbO_3$.

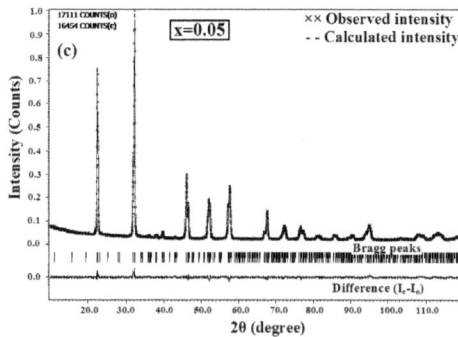

Figure 2 (c) Rietveld refinement profile for $Na_{0.95}K_{0.05}NbO_3$.

Table 1 Refined structural parameters from Rietveld refinement of $Na_{1-x}K_xNbO_3$ (x=0.01, 0.02, 0.03).

Parameters	x=0.01	x=0.03	x=0.05
a (Å)	5.4375 (5)	5.4436 (9)	5.4479 (1)
b (Å)	5.5359 (9)	5.5470 (8)	5.5518 (6)
c (Å)	15.3344 (3)	15.3365 (3)	15.3654 (3)
α=β=γ (°)	90 °	90 °	90 °
Volume ($Å^3$)	473.92(2)	479.88 (10)	480.04 (9)
Density (gm/cc)	4.5970 (5)	4.5577 (10)	4.5473 (9)
Rp (%)	3.88	3.69	3.29
Robs (%)	1.82	2.46	2.05
GOF	1.28	1.27	1.17
F(000)	609	610	611

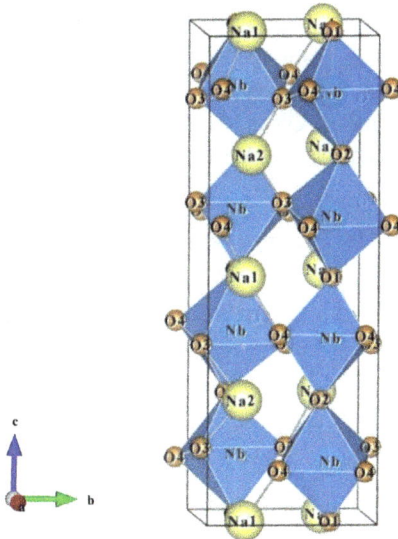

Figure 3 Polyhedral structure of $Na_{0.99}K_{0.01}NbO_3$.

3.3 CHARGE DENSITY ANALYSIS

The powder X-ray diffraction profiles of the prepared samples are refined using the JANA 2006 [17] software. Then, the charge density distribution in the unit cell for the chosen systems is determined via the maximum entropy method [16] (MEM). MEM (maximum entropy method) is an exact tool to study the electron density distribution due to its resolution. The bonding nature and the distribution of electrons in the bonding region can be clearly visualized using this technique. For the quantitative enumeration of MEM charge density a software package PRIMA (PRactice of Iterative MEM Analysis) [19] is used. Mapping of the charge density distribution is done using the visualization software package VESTA (Visualization for Electronic and STructural Analysis) [20]. In our calculations, the unit cell was divided into 64×64×162 pixels along each crystallographic axes. Prior electron density at each pixel is fixed uniformly as F_{000}/a_0^3 where F_{000} is the total number of electrons in the unit cell and a_0 is the cell parameter.

Three-dimensional electron densities imposed on the structure of $Na_{1-x}K_xNbO_3$ (x=0.01, 0.03 and 0.05) in the form of iso-surface in the unit cell are given in Figure 4 (a-c). Two-dimensional space charge distribution on (100) plane of NKN shows the charges between the atoms with the contour level between 0.0 and 1.0 $e\text{Å}^{-3}$ with an interval of 0.05 $e\text{Å}^{-3}$ (Figure 5 (b-d)). The numerical values of the MEM electron densities from one-dimensional electron density profiles between the atoms (Na-O, Nb-O) are given in table 2. Bond critical point (BCP) quantifies the presence of the space charges at the mid-bond region (Table 2). We found that the Na-O and Nb-O bond is a ionic bond in NKN (Figure 6 a and b), which has a minimum charge density at the middle of the bond which is tabulated in table 2.

Table 2 Bond lengths and mid bond electron densities for Na-O & Nb-O bondings for $Na_{1-x}K_xNbO_3$(x=0.01, 0.03, 0.05).

Samples	Bonding			
	Na-O		Nb-O	
	Bond length (Å)	Mid bond electron density (e/Å^3)	Bond length (Å)	Mid bond electron density (e/Å^3)
x=0.01	2.7842	0.2481	1.9216	0.8192
x=0.03	2.7996	0.1419	19454	0.4826
x=0.05	2.8331	0.1970	1.9610	0.4904

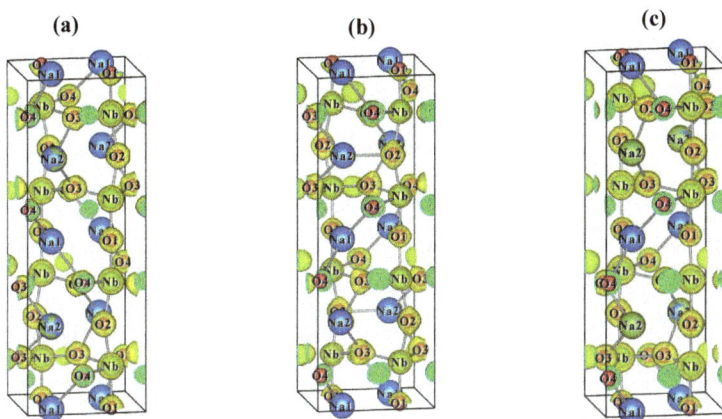

Figure 4 3-dimensional electron density for Na₁₋ₓKₓNbO₃, a) x=0.01 b) 0.03 c) 0.05.

$Figure\ 4\ 3\text{-}dimensional\ electron\ density\ for\ Na_{1-x}K_xNbO_3,\ a)\ x{=}0.01\ b)\ 0.03\ c)\ 0.05.$

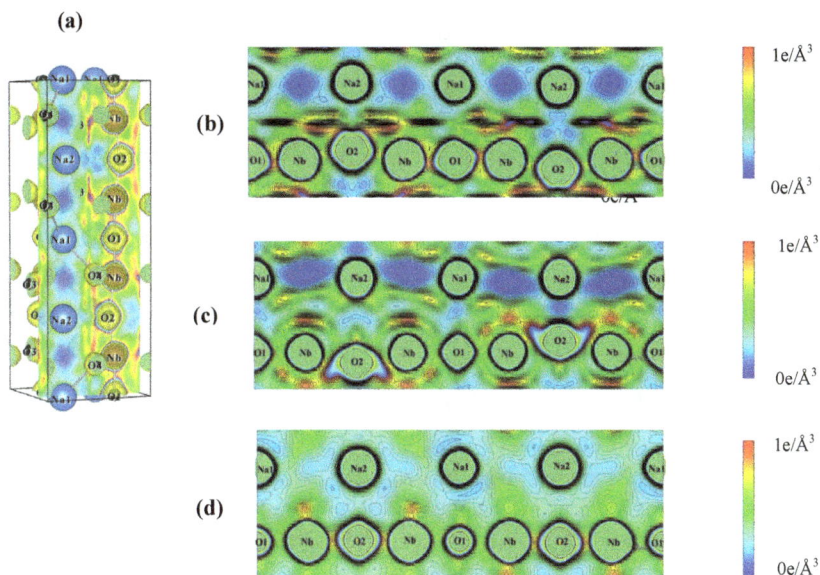

Figure 5 (a) 3-dimensional unit cell of NKN with (100) plane shaded. 2-dimensinal electron density distributions of $Na_{1-x}K_xO_3$ for b) x=0.01 c) x= 0.03 d) x=0.05 on the (100) plane.

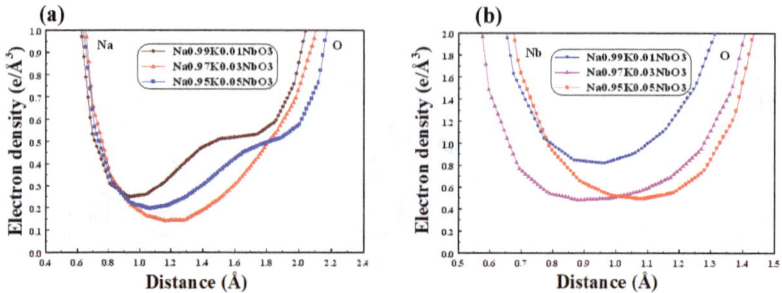

Figure 6 One dimensional electron density profiles for $Na_{1-x}K_xNbO_3(x=0.01, 0.03, 0.05)$ a) Na-O bond b) Nb-O bond.

3.4   UV-VISIBLE ANALYSIS

Figure 7 (a and b) show the UV-Visible absorption spectra for all prepared samples $Na_{1-x}K_xNbO_3$ (x=0.01, 0.03 and 0.05). Figure 7(a) shows the clear picture of absorption between 200 and 400 nm. The wave length at which the absorption occurs, increases for the samples from x=0.01, 0.03 and 0.05 respectively. The absorption spectra show that the UV-Visible absorption is decreased with the additional K^+ doping. An estimate of the optical bandgap is obtained using the following equation, $(\alpha h v)^2 = A (h v - E_g)$ [21]. Figure 7(b) shows UV-Visible spectra of $Na_{1-x}K_xNbO_3$. The optical bandgap energy of $Na_{1-x}K_xNbO_3$ is decreased with the increase of potassium content. During the substitution of K^+ ions, the carrier concentration is increased and consequently the bandgap is reduced, which could be due to the increase of particle size. The value of E_g decreases from 3.4968 eV, 3.4174 eV and 3.3466 eV for the sample x=0.01, 0.03 and 0.05 respectively. The decrease of band gap can be attributed to the increase of the particle size.

3.5   SCANNING ELECTRON MICROSCOPY AND EDS

Figure 8 (a-c) shows the scanning electron microscopy images of sintered powders of all samples. From the SEM micrographs, the sizes of the powder particle can be found. The average particle sizes are found to be 590 nm, 610 nm and 655 nm for x=0.01, 0.03 and 0.05 respectively. The particle size is found to increase with increase of K^+ content. This could be attributed to the larger ionic radii of K^+ (1.33Å) as compared to Na+ (1.02Å). Energy dispersive X-ray (EDS) analysis of $Na_{1-x}K_xNbO_3$ (x=0.01, 0.03 and 0.05) samples are shown in Figure 8 (d-f). EDS results show that the doped ions are present in the respective samples in expected proportions (Table 3). No additional impurities are detected in the EDS spectrum.

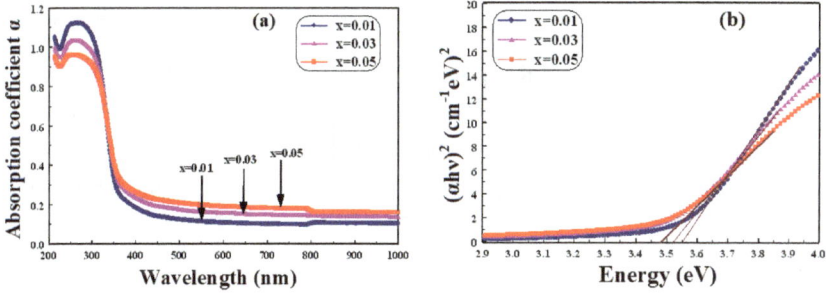

Figure 7 (a) Absorption spectra and (b) UV-Visible spectra for $Na_{1-x}K_xNbO_3$.

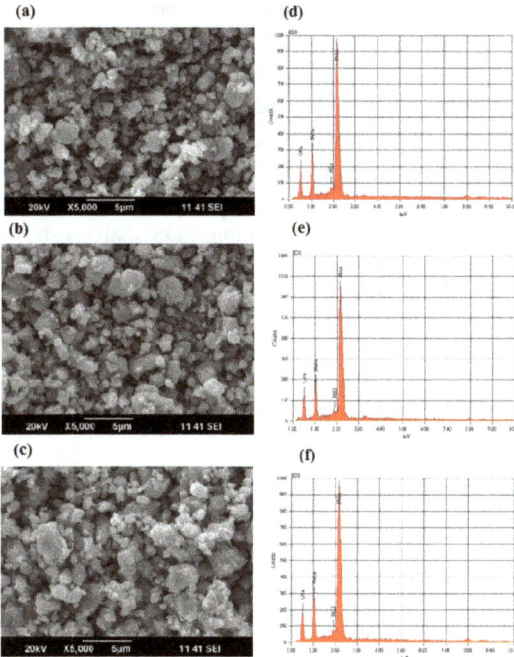

Figure 8 SEM micrographs of the prepared $Na_{1-x}K_xNbO_3$, a) x=0.01 b) x=0.03 c) x=0.05 ceramics with their elemental composition found using EDAX, (d) x=0.01 (e) x=0.03 (f) x=0.05.

Table 3 Elemental composition of the $Na_{1-x}K_xNbO_3$ ceramics determined by EDS analysis.

Compositions	Atomic percent (%)				Weight percent (%)			
	Na	K	Nb	O	Na	K	Nb	O
x=0.01	14.77	0.34	68.01	16.88	4.88	0.28	90.96	3.88
x=0.03	15.44	0.38	67.54	16.64	5.12	0.29	90.75	3.84
x=0.05	14.77	0.41	67.94	16.88	5.12	0.32	90.92	3.88

CONCLUSION

$Na_{1-x}K_xNbO_3$ (x=0.01, 0.03 and 0.05) ceramics were prepared with the solid state reaction method. The structural, optical and morphology properties of $Na_{1-x}K_xNbO_3$ ceramic samples have been studied. X-ray powder diffraction confirms the formation of orthorhombic single phase crystalline structure without any additional phase content. All XRD patterns were analyzed by using the profile refinement technique revealing the orthorhombic structure. The orthorhombic structure has been constructed using the refinement information. The charge density analysis was done using the refined structural values. UV-Visible absorption spectra indicated that the decrease of E_{gap} values is attributed to increase of potassium concentration. EDS analysis shows that the quantities of the elements present in the sample increase with composition x.

REFERENCES

[1]   Y. Saito, H. Takao, T. Tani, T. Nonoyama, K. Takatori, T. Homma, et.al. Lead-free piezoceramics, Nature 432 (2004) 84–7.
       http://dx.doi.org/10.1038/nature03028

[2]   E.K. Akdogan, K. Kerman, M. Abazari, A. Safari, Origin of high piezoelectric activity in ferroelectric (K0.44Na0.52Li0.04)–(Nb0.84Ta0.1Sb0.06)O3 ceramics, Appl. Phys. Lett. (2008) 92.

[3]   K. Wang, F.-Z. Yao, W. Jo, D. Gobeljic, V.V. Shvartsman, D.C. Lupascu, etal. Temperature-insensitive (K, Na)NbO3-based lead-free piezo actuator ceramics, Adv. Funct. Mater. 23 (2013) 4 079–86.

[4]   M.-S. Kim, S. Jeon, D.-S. Lee, S.-J. Jeong, J.-S. Song, Lead-free NKN-5LT piezoelectric materials for multilayer ceramic actuator, J. Electro. Ceram. 23 (2009) 372–5.

http://dx.doi.org/10.1007/s10832-008-9470-x

[5] S. Kawada, M. Kimura, Y. Higuchi, H. Takagi, (K,Na)NbO3-based multilayer piezoelectric ceramics with nickel inner electrodes. Appl. Phys. Express. (2009) 2. *http://dx.doi.org/10.1143/apex.2.111401*

[6] J. Jin, D. Wan, Y. Yang, Q. Li, M. Zha, A linear ultrasonic motor using (K0.5Na0.5) NbO3 based lead-free piezo electric ceramics. Sens. Actuator A: Phys. 165 (2011) 410–4. *http://dx.doi.org/10.1016/j.sna.2010.10.017*

[7] S. In-Tae, K. In-Young, C. You-Jeong, C. Jae-Hong, N. Sahn, S. Tae-H yun, et al. Piezoelectric properties of CuO-added (Na0.5K0.5)NbO3 ceramic multilayers. J. Eur. Ceram. Soc. 32 (2012) 1085–90. *http://dx.doi.org/10.1016/j.jeurceramsoc.2011.11.020*

[8] K. Motoo, F. Arai, T. Fukuda, M. Matsubara, K. Kikuta, T. Yamaguchi, et al. Touch sensor for micro manipulation with pipette using lead-free (K, Na)(Nb, Ta)O3 piezoelectric ceramics. J. Appl. Phys. (2005) 98.

[9] M. Guo, K.H. Lam, D.M. Lin, S. Wang, K.W. Kwok, H.L.W. Chan, et al. A Rosen-type piezoelectric transformer employing lead-free K0.5Na0.5NbO3 ceramics, J. Mater. Sci. 43 (2008) 709–14. *http://dx.doi.org/10.1007/s10853-007-2199-0*

[10] E. Cross, Materials Science-lead-free at last, Nature 432 (2004) 24–5. *http://dx.doi.org/10.1038/nature03142*

[11] G. Shirane, R. Newnham, R. Pepinsky, Dielectric properties and phase transitions of NaNbO3 and (Na, K)NbO3, Physical Review 96 (1954) 581–588. *http://dx.doi.org/10.1103/PhysRev.96.581*

[12] R.E. Jaeger, L. Egerton, Hot pressing of potassium-sodium niobates, Journal of the American Ceramic Society, 45 (1962) 209-213. *http://dx.doi.org/10.1111/j.1151-2916.1962.tb11127.x*

[13] Y. Saito, H. Takao, T. Tani, T. Nonoyama, K. Takatori, T. Homma, T. Nagaya, M. Nakamura, Lead-free piezoceramics, Nature 432 (2004) 84-87. *http://dx.doi.org/10.1038/nature03028*

[14] L. Egerton, D.M. Dillom, Piezoelectric and Dielectric Properties of Ceramics in the System Potassium-Sodium Niobate, J. Am. Ceram. Soc. 42 (1959) 438. *http://dx.doi.org/10.1111/j.1151-2916.1959.tb12971.x*

[15] H.M. Rietveld, A Profile Refinement Method for Nuclear and Magnetic, J. Appl. Crystallogr. 2 (1969) 65.

http://dx.doi.org/10.1107/S0021889869006558

[16]   D.M. Collins DM (1982) Electron density images from imperfect data by iterative entropy maximization, Nature **49 298**.

[17]   V. Petricek, M. Dusek and L. Palatinus, (2006) Jana, The crystallographic computing system (Institute of Physics), Praha, Czech Republic.

[18]   R. D. Shannon, Revised effective ionic radii and systematic studies of interatomic distances in halides and chalcogenides. Acta Cryst. A32 (1976) 751-767. *http://dx.doi.org/10.1107/S0567739476001551*

[19]   K. Momma and F. Izumi, Commission on Crystallogr. Comput., IUCr Newslett 7 (2006) 106.

[20]   F. Izumi, R.A. Dilanian, IUCr Newslett 32 (2005) 59.

[21]   J.I. Pancove, (1971). Optical processes in semiconductors. Englewood Cliffs, NJ, USA: Prentice Hall.

CHAPTER 3

Inter Bond Experimental Electron Density in Magnesium Ferrite Ceramic (MgFe$_2$O$_4$) Through XRD

M.J. Viswanath[1], M. Aysha kani[1], S.V. Meenakshi[2] and R. Saravanan[1,*]

[1]Research centre and PG Department of Physics, The Madura College, Madurai-625 011, Tamil Nadu, India

[2]Department of Physics, Sri Meenakshi Government Arts College for women, Madurai-625 002, Tamil Nadu, India

Email: raiswa003@gmail.com; aysha.m252@gmail.com; svmeenu74@yahoo.in; saragow@gmail.com*

Abstract

The magnesium ferrite (MgFe$_2$O$_4$) sample is prepared with the solid state reaction method. It was characterized by Powder-XRD, UV-visible and SEM. The average crystallite sizes are calculated by the Debye Scherrer's formula in the range of micrometers. From the XRD data the electron density is evaluated using the MEM (Maximum Entropy Method) analysis. UV- visible spectroscopy was used to calculate the band gap value of the material and the particle size is measured through SEM.

Keywords

Spinel Cubic, Electron Density, Bonding Nature, MEM, Direct Band Gap, Rietveld Refinement, SEM & UV-Visible.

Contents

1.   INTRODUCTION

In recent years magnetic materials have attracted much attention because of their unique and novel properties and a wide variety of potential applications, such as information storage, bioprocessing, color imaging and magnetic refrigeration [1]. Magnesium ferrite is one of the most important ferrites. Since it is a soft magnetic n-type semiconducting material it has a wide energy gap. Magnesium ferrite has been found to exhibit some unusual magnetic properties, such as super paramagnetism and a non-collinear ordering of the magnetic moments of Fe3+ ions, known as spin canting [2-4]. It is spinel cubic in structure. The spinel structure (sometimes called garnet structure) is named after the mineral spinel ($MgFe_2O_4$); the general composition is AB_2O_4. It is essentially cubic, with the O⁻ ions forming a FCC lattice. The cations (usually metals) occupy 1/8 of the tetrahedral sites and 1/2 of the octahedral sites and there are 32 O⁻ ions in the unit cell [5]. We report here a structural study of $MgFe_2O_4$ prepared by hand grinding. We find that its atomic arrangement may be well described in terms of a spinel type structure.

2.   SAMPLE PREPARATION

The raw materials Magnesium Oxide (MgO 99%) & Iron II Oxide (Fe_2O_3 99.99%) were taken in stoichiometric ratio and ground using an agate mortar for 11 hours. The prepared sample was sintered at 1200 °C for 5 hours until it turned into a crystalline magnesium ferrite. The thermally treated sample was characterized by powder XRD, UV-Visible and SEM.

2.1   POWDER X-RAY DIFFRACTION

The powder XRD experiments were carried out on a Philips X'Pert diffractometer using cu kα radiation. The diffraction patterns were collected in the 2θ range from 25° to 120° with step size of 0.02°. It is an analytical technique used for the phase identification of a crystalline material. It provides information on the unit cell dimensions, atomic arrangement, size and shape of a unit cell [6]. The Rietveld refinement is the standard tool devised by Hugo Rietveld [7] to be used in the characterization of crystalline materials. The Rietveld refinement is performed using the improvised version of the JANA 2006 software. In the present work, cell parameters, structural parameters, occupancy of atom were refined using JANA 2006 [8]. The cell parameters and structural parameters are given in table 1. The Rietveld refinement profile and the positions of the Bragg peaks are shown in figure 1.

Table 1: Structural parameters of MgFe$_2$O$_4$.

Parameter	Value
a=b=c (Å)	8.397
Cell volume (Å^3)	586.7302
Density (gm/cc)	4.9535
R$_{obs}$ (%)	5.61
$_w$R$_{obs}$ (%)	5.33
R$_p$ (%)	1.57
$_w$R$_p$ (%)	2.03
GOF	1.13

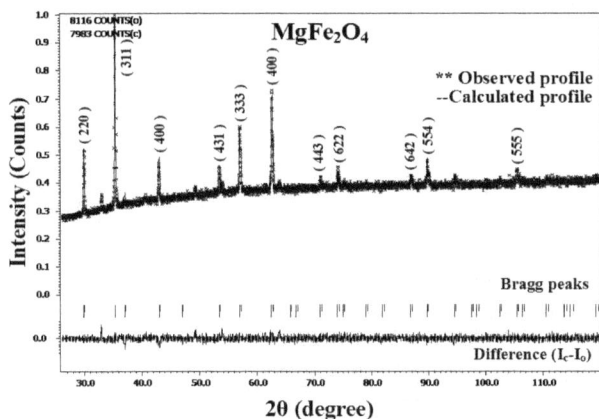

Fig. 1 Rietveld refinement profile of MgFe$_2$O$_4$.

Table 2. Structure factor for MgFe$_2$O$_4$ obtained from Rietveld refinement.

h k l	Fo	Fc	σ(Fo)
2 0 2	150.69	153.01	1.992
1 1 3	227.45	226.39	2.431
2 2 2	74.36	109.41	4.802
0 0 4	259.50	284.64	4.224
3 1 3	34.25	37.13	8.905
2 2 4	139.56	130.33	3.361
3 3 3	200.26	198.31	2.657
1 1 5	215.06	212.97	2.853
4 0 4	453.62	448.86	5.426
3 1 5	29.20	18.03	8.005
4 2 4	17.25	8.93	11.312
2 0 6	110.41	115.14	7.672
3 3 5	162.91	167.98	5.411
2 2 6	79.52	81.21	5.109
4 4 4	139.22	149.89	-19.036
1 1 7	42.64	41.62	15.623
2 2 8	92.57	104.55	9.363
4 2 6	100.25	114.44	5.499
3 1 7	144.34	152.92	3.067
5 3 5	151.61	160.63	3.222
0 0 8	331.62	326.61	18.415
3 3 7	51.38	44.48	22.725
4 4 6	75.01	89.61	20.532
6 4 6	101.50	128.81	12.052
6 0 6	104.57	118.11	10.577
5 5 5	125.80	123.29	5.171
5 1 7	129.37	129.20	5.129
6 2 6	47.39	52.32	10.574
4 0 8	114.70	90.78	20.837
1 1 9	53.19	35.33	19.737
5 3 7	27.31	18.41	10.288
4 2 8	58.63	68.05	15.909

2.2   MAXIMUM ENTROPY METHOD

MEM is an exact tool that can yield a high-resolution density distribution from a limited number of diffraction data. PRIMA [9, 10] & VESTA [11] are MEM analysis programs which can be used to determine the electron densities in the inner atomic region for example, the bonding region [12].The MEM refinements were carried out by dividing the unit cell into into 48x48x48 pixels. The uniform prior density was obtained by dividing the total number of electrons by the volume of the unit cell. The volume of a unit cell is 586.776. The MEM parameters are given in table 3.

Table 3. Parameters from MEM refinement.

Parameters	$MgFe_2O_4$
Number of cycles	2046
Lagrange parameter (λ)	0.015225
(F_{000})	842
R_{MEM} (%)	0.021055
$_wR_{MEM}$ (%)	0.001723

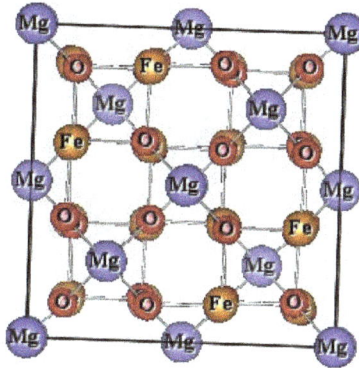

Fig. 2(a) 3D electron density of $MgFe_2O_4$ in the unit cell.

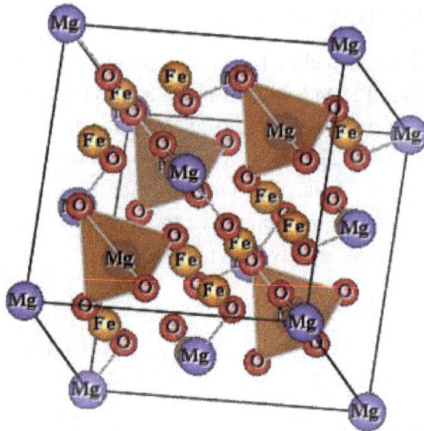

Fig. 2(b) 3D tetrahedral sites of $MgFe_2O_4$ in the unit cell.

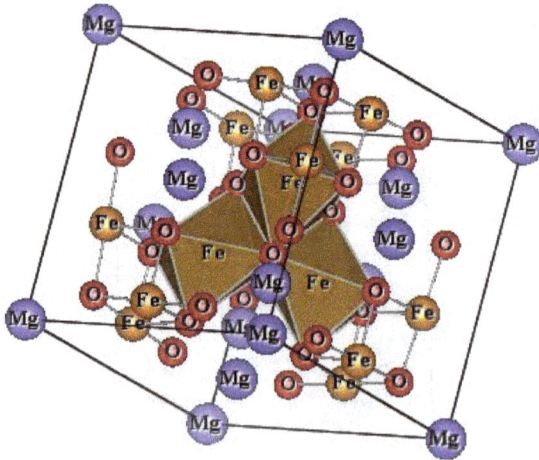

Fig. 2(c) 3D octahedral sites of $MgFe_2O_4$ in the unit cell.

In figure 2(a) the unit cell of magnesium ferrite is shown using the PRIMA & VESTA software. The bonding between the Mg & O atoms and the Fe & O atoms in the unit cell is also shown in the figure. The spinel cubic structure is essentially cubic with the formation of tetrahedral sites and octahedral sites. In figure 2(b) we can see that four oxygen atoms surround the Magnesium atom thus forming the tetrahedral sites. Similarly in figure 2(c) we can see that the Fe atom is surrounded by six O atoms forming the octahedral sites.

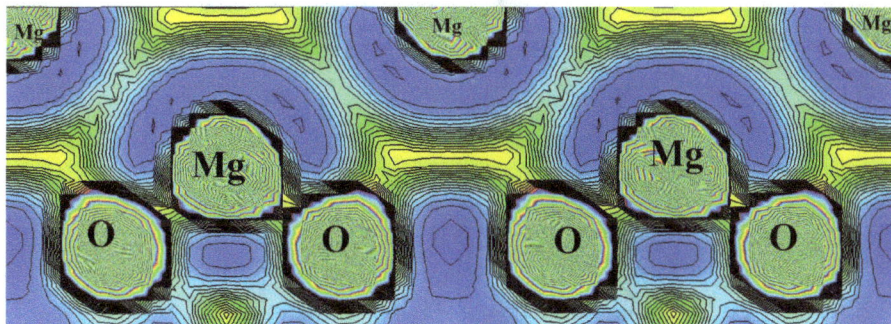

Fig. 3(a) 2D electron density map of $MgFe_2O_4$ on (011) plane.

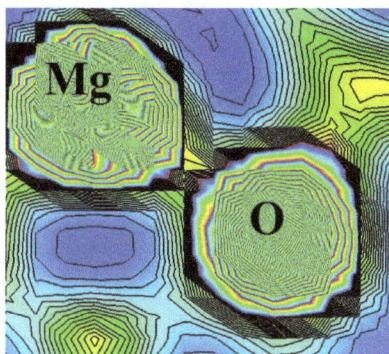

Fig. 3(b) Enlarged view of Mg-O bond.

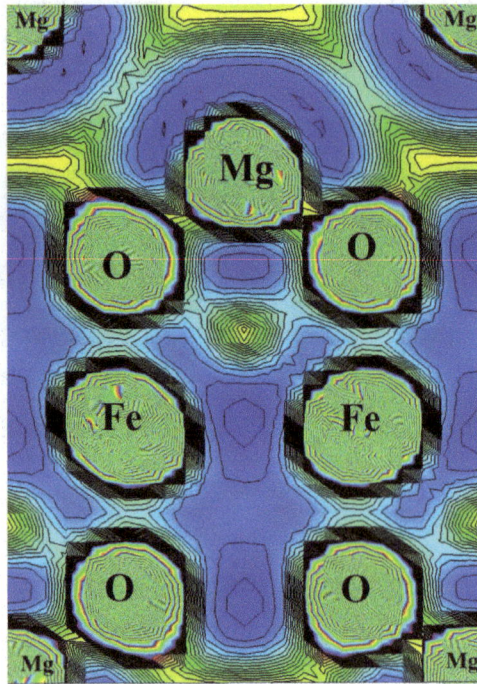

Fig. 3(c) 2D electron density map of $MgFe_2O_4$ on (011) plane.

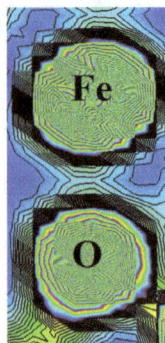

Fig. 3(d) Enlarged view of Fe-O bond.

The 2D electron density of $MgFe_2O_4$ on (011) plane drawn in the x direction with x=0.5 (min) is shown in figure 3(a). Figure 3(c) shows the 2D electron density of $MgFe_2O_4$ on the (011) plane in the y direction with y=0.5 (min).

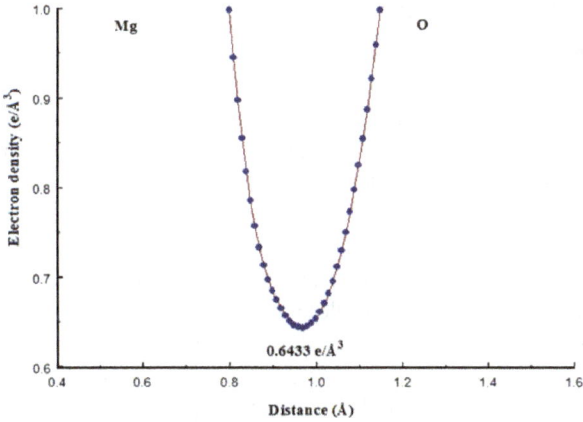

Fig. 4(a) 1D electron density profile of $MgFe_2O_4$.

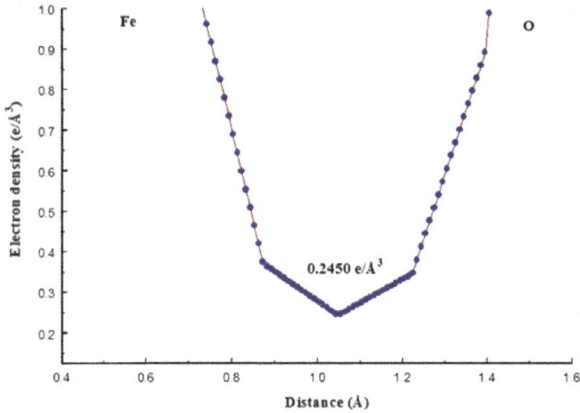

Fig. 4(b) 1D electron density profile of $MgFe_2O_4$.

Fig. 4(a) & 4(b) shows the one dimensional electron density distribution of $MgFe_2O_4$ on the (011) plane. From fig. 4(a) the bonding between Mg and O atom seems to be covalent. The mid bond electron density of Mg-O bond is $0.6433e/Å^3$. In fig. 4(b) it is shown that the bonding between the Fe and the O atom seems to be ionic. The mid bond electron density of the Fe-O bond is $0.2450e/Å^3$.

2.3   UV-VISIBLE ANALYSIS

The UV-visible analysis is an absorption spectroscopy or reflectance spectroscopy in the ultraviolet-visible spectral region. This means it uses light in the visible and adjacent ranges. The colors of the chemicals involved are affected by the absorption or reflectance in the visible range. Molecules are undergoing electronic transitions in this region of electromagnetic radiation. In this technique, absorption measures transition from the ground state to the excited state [13]. The basic principle of UV absorption is that the molecule absorbs energy in the form of electromagnetic radiation to excite the electron to higher anti bonding molecular orbitals. The easier the electrons are excited the longer the light can be absorbed. Different molecules absorb radiation of different wavelengths. An absorption spectrum will show a number of absorption bands corresponding to structural groups within the molecule. In order to study the band gap of a material, a light with variable photon energy is used near the band gap energy. The photon energy at which the transition between absorbing and non-absorbing behavior takes place will correspond to the band gap energy [14]. From the UV-visible data, $MgFe_2O_4$ is found to be a semiconductor and its band gap value is about 2.34 ev. The optical absorption coefficient (α) of a semiconductor will depend on the wavelength because of the relationship between absorption and the semiconductor band gap. Since $MgFe_2O_4$ obeys the relation $\alpha=K_d (E_p - E_g)^{1/2}$ it is a direct band gap semiconductor. Where α is the optical absorption coefficient, E_P is the photon energy and E_g is the energy required to break a covalent bond. There is only one proportional constant, because direct band gap materials do not involve phonons (lattice vibrations) in electron band-to-band transitions [15].

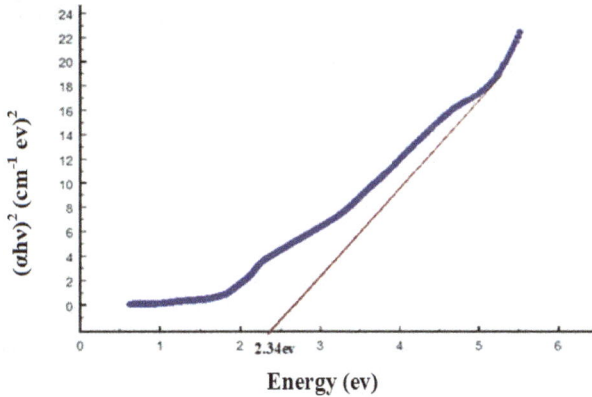

Fig. 5 Band gap profile of $MgFe_2O_4$.

2.4 SEM ANALYSIS

Scanning electron microscope is a versatile, non-destructive technique used to find the particle size, morphology and composition of materials. SEM can produce clearer image with spatial resolution up to nanometer range [16]. SEM analyses were carried out for various magnifications one of which is shown in figure. The average size of the particle detected in this figure can be compared with the size calculated with XRD. The size of $MgFe_2O_4$ is analyzed using SEM. The average particle size of $MgFe_2O_4$ compared with the size calculated with XRD is given in table 4. Using Debye Scherrer's formula the grain size of $MgFe_2O_4$ was calculated [17].

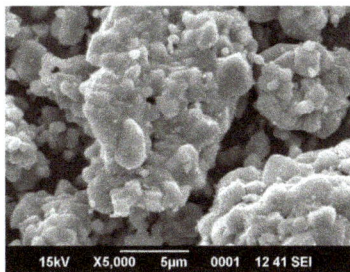

Fig. 6 SEM picture of $MgFe_2O_4$ with a magnification ×5000.

Table 4. The average particle size from SEM & XRD.

System	Particle size (SEM)	Grain size (XRD)
$MgFe_2O_4$	1.27μm	0.0348μm

3. CONCLUSION

The XRD patterns confirmed that $MgFe_2O_4$ possess a spinel cubic structure. The electron density has been analyzed with XRD. From the UV-visible data, $MgFe_2O_4$ is found to be a semiconductor and its band gap value is about 2.34 ev. SEM results show that the average crystallite size of the powders is in the micrometer range. Further procedures are planned and the results will be reported in future publications.

ACKNOWLEDGEMENTS

The authors wish to thank the authorities of Madura College for the basic infrastructure provided for active research in the Dept. of Physics. This work was not supported by any funding agencies.

REFERENCES

[1]   D. L. Leslie-Pelecky, R. D. Rieke: Chem. Magnetic Properties of Nanostructured Materials 8 (1996), 1770-1783.

[2]   R. J. Willey, P. Noirclerc & G. Busca: Chem. Ningbo Institute of Materials Technology and Engineering (NIMTE). Commun. 123 (1993), 1-16.

[3]   V. Šepelák, D. Baabe, D. Mienert, F. J. Litterst, K. D. Becker, "Enhanced Magnetisation in Nanocrystalline High-Energy Milled MgFe2O4", Srcipta Materialia 48 (2003) 961-966
       http://dx.doi.org/10.1016/S1359-6462(02)00600-0

[4]   V. Šepelák, M. Menzel, K. D. Becker, F. Krumeich, "Mechanochemical Reduction of Magnesium Ferrite", J. Phys. Chem. B 106 (2002) 6672-6678
       http://dx.doi.org/10.1021/jp020270z

[5]   Daliya S. Mathew, Ruey-Shin Juang: Chemical Engineering Journal, 129 (2007), 51-65
       http://dx.doi.org/10.1016/j.cej.2006.11.001

[6]   B.D. Cullity: Elements of X-ray diffraction. Addison-Wesley series in Metallurgy & materials, (1978)

[7]   H.M. Rietveld: J. Appl. Crystallogr. 2 (1969) 65
        http://dx.doi.org/10.1107/S0021889869006558

[8]   V. Petricek, M. Dus ek, L. Palatinus, in: JANA 2000, The Crystallographic Computing System, Institute of Physics, Academy of Sciences of the Czech Republic, Praha, 2000.

[9]   A.D. Ruben, I. Fujio: Super-fast Program PRIMA for the Maximum Entropy Method, Advanced materials Laboratory, National Institute for Materials Science, Ibaraki, Japan (2004).

[10]   F. Izumi, R.A. Dilanian: Recent Research Developments in Physics, Part II, 3, Transworld, Research Network, Trivandrum, 2002, pp. 699-726.

[11]   K. Mooma, F. Izumi: J. Appl. Crystallogr. 41 (2008) 653.
        http://dx.doi.org/10.1107/S0021889808012016

[12]   D.M. Collins: Nature 49 (1982) 298.

[13]   S.O Kasap: Electrical Engineering Materials and Devices. Irwin (1997) Chapters 5.7

[14]   B.Streetman: Solid-State Electronic Devices. Prentice-Hall. (1995), Chapter 4.1

[15]   O. Stenzel: The Physics of Thin Film Optical Spectra: An Introduction. Springer. (2005),214.

[16]   L.B. Fraigi, D.G. Lamas and N.E. Walsoe de Reca, Material Letters 47 (2001) 262-266
        http://dx.doi.org/10.1016/S0167-577X(00)00246-9

[17]   R. Saravanan: GRAIN software available at http://www.saraxraygroup.net/

CHAPTER 4

Effects of Cations Substitution on Structural and Magnetic Properties of LaCrO$_3$ Ceramic Perovskites

N. Thenmozhi[a], R. Saravanan[b*], Yen-Pei Fu[c]

[a]PG and Research Department of Physics, NMSSVN College, Nagamalai, Madurai-625 019, India

[b]Research Centre and Post Graduate Department of physics, The Madura College, Madurai – 625 011, India

[c]Department of materials Science and Engineering, National Dong-Hwa University, Shou-Feng, Hualien 974, Taiwan

Email: *saragow@gmail.com; thenmozhi.n6@gmail.com

Abstract

The co-doped lanthanum chromite $(La_{0.8}Ca_{0.2})(Cr_{0.9}Co_{0.1})O_3$ was synthesized using the solid state reaction technique. The sample has been characterized by X-ray diffraction for a structural and charge density analysis. The sample was analyzed by UV-visible spectrometry for optical properties and scanning electron microscopy for surface morphology. Further, the sample was also investigated by the vibrating sample magnetometry for its magnetic properties. XRD data show that the grown sample is orthorhombic in structure with a single phase. The UV absorption spectra gives the energy band gap of the sample as 2.4450 eV. The M-H curve obtained from VSM measurements exhibit weak ferromagnetism of the grown sample.

Keywords

X-Ray Diffraction, Ceramics, Charge Density, Magnetic Property, Scanning Electron Microscopy.

Contents

1.   INTRODUCTION

Lanthanum chromite ($LaCrO_3$), a perovskite ceramic material has been studied over the past several decades due to its technological applications. This material has a high melting point (2490 °C), high mechanical strength, a high thermal expansion coefficient, good electrical conductivity at high temperatures (>800 °C) and stable chemical and physical properties in both oxidizing and reducing atmospheres [1]. Because of these properties, lanthanum chromite has been used as electrodes for magneto hydrodynamic [MHD] generators [2]. These materials are used as oxidation catalysts for soot combustion [3, 4]. Since lanthanum chromite is an oxide based material having high electrical conductivity, the material has found use in electric heaters [5]. The material is also used in high temperature NO_x sensors [6, 7]. But the alkaline earth metal doped lanthanum chromites are considered to be a major material for the preparation of interconnectors in solid oxide fuel cells (SOFC) which is a solid electrochemical device converting chemical reaction energy directly into electrical energy without combustion [8]. Since the doped lanthanum chromites exhibit mixed electronic and ionic conductivity, they are used as electrodes in SOFC. Acceptor substituted lanthanum chromite perovskite oxides are currently considered to be the most promising interconnect materials [9]. The doped lanthanum chromites are used as protective coatings on metallic interconnects [10]. In the present work, in $LaCrO_3$, Ca is doped on the La site, Co is doped on the Cr-site and the solid state reaction method is used for preparing the sample. In in this work, we report on structural, morphological, optical and magnetic properties of co-doped lanthanum chromite.

2. EXPERIMENTAL

2.1 SAMPLE PREPARATION

The ceramic perovskite sample $(La_{0.8}Ca_{0.2})(Cr_{0.9}Co_{0.1})O_3$ has been prepared by the conventional solid state reaction method. Stoichiometric amounts of lanthanum oxide $(La_2O_3$, 99.99% pure), calcium carbonate $(CaCO_3$, 99.99% pure), chromium oxide $(Cr_2O_3$, 99.99% pure), and cobalt oxide $(Co_3O_4$, 99.99% pure) were mixed with distilled water for 12 hours. After drying, the powder was ground and then calcined in air at $1000°C$ for 4 hours. The prepared powder sample was then pelletized. The pellets were dry pressed at 100MPa. Finally, these pellets were sintered out in air at $1500°$ C for 6 hours with a programmed heating rate of 5 °C/min.

2.2 CHARACTERIZATIONS

The synthesized sample has been analyzed using powder X-ray diffraction. The XRD datasets were collected at the Sophisticated Analytical Instrument Facility (SAIF), Cochin University, Cochin, using an X-ray diffractometer (Bruker AXS D8 advance) with CuKα, monochromatic incident beam (λ=1.54056 Å) with the step size of 0.02°. The UV-Visible absorption spectra for the grown sample were recorded at SAIF, Cochin (India) using a UV-Visible spectrometer Cary 5000 (Varian, Germany). The surface morphology of the sample was analyzed by scanning electron microscopy (JEOL Model JSM - 6390LV). The SEM image of the sample was also recorded at SAIF, Cochin (India). The magnetic properties of the grown sample were investigated by a vibrating sample magnetometer (Lakeshore VSM 7410). The measurements were taken at SAIF, IIT Madras, Chennai.

3. RESULTS AND DISCUSSIONS

3.1 STRUCTURAL ANALYSIS

Phase purity of the prepared $(La_{0.8}Ca_{0.2})(Cr_{0.9}Co_{0.1})O_3$ sample was examined by X ray diffraction. The XRD pattern of the grown sample $(La_{0.8}Ca_{0.2})(Cr_{0.9}Co_{0.1})O_3$ is shown in figure 1. The XRD data shows sharp peaks with narrow full width at half maximum (FWHM), which indicate the crystals are well crystallized. The baselines of XRD pattern are smooth and there are no impurity peaks shown, which authenticates the high purity of the prepared sample. All peak positions were indexed to $LaCrO_3$ (JCPDS 33-0701). The observed XRD pattern matched well with that of the orthorhombic space group (No.62). In the present work, to analyze the structural parameters of the sample, the collected XRD data was subjected to the Rietveld method [11] of refinement using Jana 2008 [12].

For the grown sample, cell parameters, peak shift, background profile shape and preferred orientation were refined from the observed XRD profile by comparing it with the theoretically generated profile. The crystal structure of the grown sample $(La_{0.8}Ca_{0.2})(Cr_{0.9}Co_{0.1})O_3$ was refined in the orthorhombic setting which belongs to the space group of Pnma with four molecules in the unit cell. The refined profile for the grown sample $(La_{0.8}Ca_{0.2})(Cr_{0.9}Co_{0.1})O_3$ is given in figure 2. In the orthorhombic setting, La and Ca atoms are fixed at Wyckoff position 4C (0.0267, 0.25, -0.004); (Cr, Co) atoms are fixed at 4b (0, 0, 0.25); O1 apex atom is at 4C (0.4905, 0.25, 0.0684) and O2 planar atom at 8d (0.2193, 0.5361, 0.2195) [13]. The orthorhombic unit cell of $(La_{0.8}Ca_{0.2})(Cr_{0.9}Co_{0.1})O_3$ by VESTA [14] is shown in figure 3. Each unit cell of $LaCrO_3$ has corner-linked octahedra CrO_6 from the centres occupied by centro symmetric Cr atoms. The corner atoms of the octahedra are oxygen atoms and the lanthanum atoms occupy the space between the octahedra [15]. The refined structural parameters and the reliability indices for the prepared sample are given in table 1. With Ca and Co dopant content, the lattice parameters and hence the volume of the perovskite have a tendency to decrease when it is compared with undoped $LaCrO_3$ (Volume of $LaCrO_3$ is $234.52Å^3$ from JCPDS # 33-0701). This decrease can be explained by the fact that the ionic radius of Ca^{2+} (1.0Å) is smaller than La^{3+} (1.16Å) and ionic radius of Co^{3+} (0.545Å) is smaller than Cr^{3+} (0.615Å) [16].

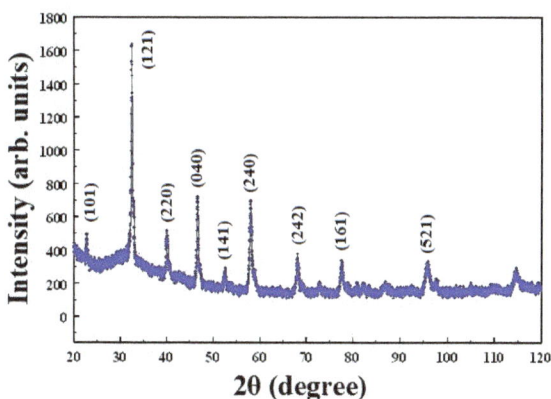

Figure 1 Observed X-ray powder diffractogram of $(La_{0.8}Ca_{0.2})(Cr_{0.9}Co_{0.1})O_3$.

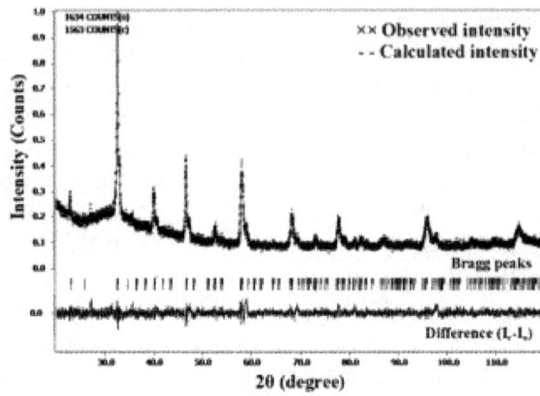

Figure 2 Fitted powder XRD profile for $(La_{0.8}Ca_{0.2})(Cr_{0.9}Co_{0.1})O_3$.

Table 1 Structural parameters for $(La_{0.8}Ca_{0.2})(Cr_{0.9}Co_{0.1})O_3$ through refinement of powder XRD data.

Parameters	Values
a (Å)	5.4982(3)
b (Å)	7.7763(8)
c (Å)	5.4756(2)
$\alpha=\beta=\gamma$ (°)	90
Volume (Å^3)	234.11(8)
Density (gm/cc)	6.23(1)
R_p (%)	6.08
R_{obs} (%)	3.92
GOF	1.15
F(000)	392

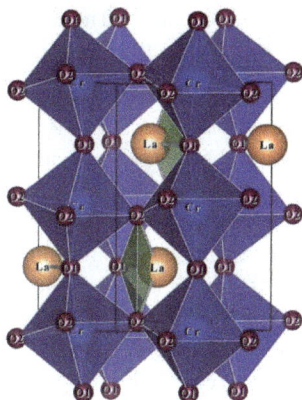

Figure 3 The unit cell of $(La_{0.8}Ca_{0.2})(Cr_{0.9}Co_{0.1})O_3$ obtained from VESTA.

3.2 CHARGE DENSITY ANALYSIS BY MEM

The distribution of charges between the atoms in the lattice can be studied with the maximum entropy method (MEM) [17] which uses structure factors extracted from the Rietveld refinement technique [11]. The MEM method [16] gives accurate pictures of distribution of charges especially in the valence region. Therefore, it is used for analyzing the bonding features and other structure based parameters. This method was implemented using the software practice iterative MEM analysis PRIMA [18]. The MEM computations for the prepared sample $(La_{0.8}Ca_{0.2})(Cr_{0.9}Co_{0.1})O_3$ was carried out using $48\times64\times48$ pixels along the a, b and c axes of the orthorhombic lattice. The resultant electron density distribution in the unit cell was visualized using the visualization software VESTA (visualization for electronic and structural analysis) [14].

For the grown sample $(La_{0.8}Ca_{0.2})(Cr_{0.9}Co_{0.1})O_3$, the three dimensional charge density distributions in the unit cell with an iso-surface level of 3.0 e/Å^3 are shown in ball and stick model structure and is presented in figure 4. The three dimensional electron density distribution of $(La_{0.8}Ca_{0.2})(Cr_{0.9}Co_{0.})O_3$ clearly gives the position of the atoms and confirms the orthorhombic phase of the grown sample. Figures 5 (a) and (c) show the 3D unit cells with (101) plane and (020) plane shaded. The two dimensional electron density distribution of La and O2 atoms are drawn in the range 0-1.0 e/Å^3 with interval 0.04 e/Å^3 for the (101) plane (figure 5(b)) and for Cr and O2 in the same range for (020) plane (figure 5 (d)). The 2-D map (figure 5(b)) authenticates that there is no sharing of electrons along the bonding region of La and O2 atoms. This confirms that the La-O2 bond is ionic with a partial covalent character. The 2-D map for the plane (020) (figure 5(d)) shows the increase of charges in the bonding region of Cr and O2 atoms, confirming that the bond Cr-O2 is covalent with partial ionic character.

To quantify these results, we have plotted one dimensional electron density profiles for the two bonds La-O2 and Cr-O2 which are shown in figures 7 and 8. The bond length and the mid bond density values are given in table 2. In table 2, it is seen that, for the La-O2 bond, the mid bond density value is 0.5248 e/Å³ indicating that the La-O2 bond is less ionic. Again in table 2, it is made clear that the mid bond density value is 0.5049 e/Å³ for Cr-O₂ bond which leads to the decrease in ionic behavior. The bond length values for La-O2 and Cr-O2 agree well with the reported values [19, 20]. The covalent character of Cr-O2 bond is attributed to the electronic conductivity of the sample which leads to the application of SOFC. The electronic conductivity of lanthanum chromite depends on the composition of the doped cations.

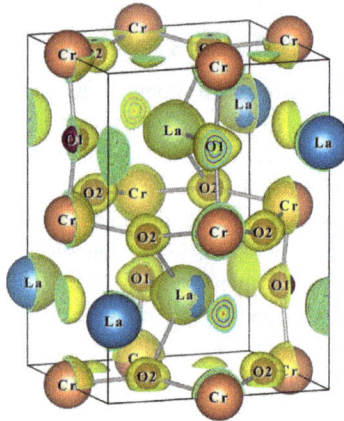

Figure 4 Three dimensional electron density isosurfaces for (La$_{0.8}$Ca$_{0.2}$)(Cr$_{0.9}$Co$_{0.1}$)O$_3$.

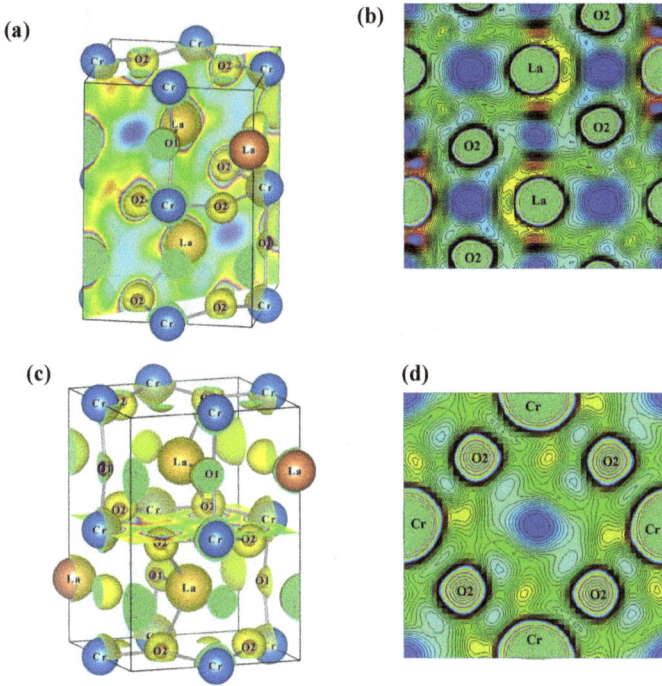

Figure 5 3D unit cell of $(La_{0.8}Ca_{0.2})(Cr_{0.9}Co_{0.1})O_3$ with a) (101) plane and c) (020) plane shaded. Two dimensional electron density distribution on b) (101) and d) (020) planes for $(La_{0.8}Ca_{0.2})(Cr_{0.9}Co_{0.1})O_3$.

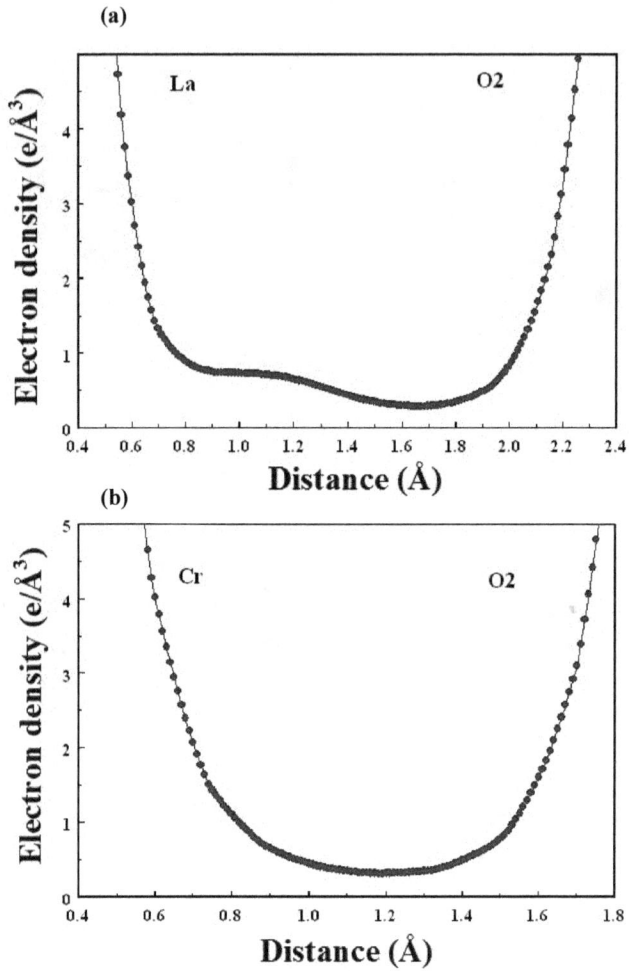

Figure 6 One dimensional electron density profiles along a) La-O2 and b) Cr-O2 bonds for $(La_{0.8}Ca_{0.2})(Cr_{0.9}Co_{0.1})O_3$.

Table 2 Bond lengths and mid bond electron densities for La-O2 and Cr-O2 bonds for $(La_{0.8}Ca_{0.2})(Cr_{0.9}Co_{0.1})O_3$.

La-O2		Cr-O2	
Bond length (Å)	Mid bond electron density (e/Å³)	Bond length (Å)	Mid bond electron density (e/Å³)
2.4819	0.5248	1.9906	0.5049

3.3   UV-VISIBLE ANALYSIS

The optical band gap of the sample was found using the UV-Visible absorption spectra, which was taken in the range from 200 to 2000 nm. The absorption coefficient and the energy band gap of the material are related by Tauc's equation [21] $\alpha h\nu = A (h\nu - E_g)^n$, where A is a constant, α is absorption coefficient, $h\nu$ is the photon energy, E_g is the energy band gap, using $n=1/2$ for direct band gap materials and $n=2$ for indirect band gap materials. Using Tauc's relation a graph is drawn with the energy value on the x-axis and $(\alpha h\nu)^2$ on the y-axis. The Tauc's region is extrapolated to $(\alpha h\nu)^2=0$, to obtain the band gap. The $((\alpha h\nu)^2$ vs $h\nu)$ plot for the grown sample is shown in figure 7 and from the plot, the direct energy band gap value obtained is 2.4450 eV. This E_g value agrees well with the reported value [22].

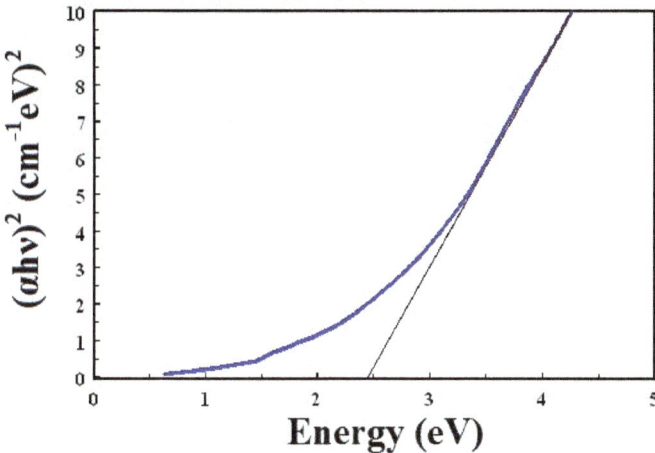

Figure 7 UV-Visible plot for $(La_{0.8}Ca_{0.2})(Cr_{0.9}Co_{0.1})O_3$.

3.4   MICROSTRUCTURE ANALYSIS

The crystallite size of the grown sample $(La_{0.8}Ca_{0.2})(Cr_{0.9}Co_{0.1})O_3$ was estimated from the XRD powder data using the GRAIN software [23] which employs the Scherrer formula [24] $t = 0.9\lambda/\beta \cos\theta$, where t is the grain size (size of the coherently diffracting domain), λ is the wavelength of the X-ray, β is the full width at half maximum and θ is the Bragg angle of reflection. The Grain size of the prepared sample was in the range of 13 nm to 17 nm which is in agreement with the previously reported values [25]. The SEM images of the sample were recorded for various magnifications (x1500 x5000 x10000), and the image for the magnification of ×10,000 is shown in figure 8. In the SEM image, it is visible that the particles are finely distributed without much agglomeration and the particle sizes are in the range of sub micrometers [26].

Figure 8 SEM image for $(La_{0.8}Ca_{0.2})(Cr_{0.9}Co_{0.1})O_3$.

3.5   MAGNETIC PROPERTIES

Rare earth chromites in general possess the property of ferroelectric or antiferromagnetic multiferroics. Lanthanum chromite is a G type antiferromagnetic below 290 K and is a poor electrical conductor at room temperature [27]. Studies on Ca doped $LaCrO_3$ showed that, with increasing calcium doping concentration, the Neel temperature T_N decreases from 290 K to 160 K and a weak ferromagnetic state with large coercive fields is observed. This weak ferromagnetic moment is due to canting of the antiferromagnetically ordered Cr moments [28]. The M-H loop recorded for the grown sample at room temperature is shown in figure 9. The magnetic parameters such as retentivity, coercivity and saturation magnetization are listed in table 3. The non-saturation behavior of M-H curve with observed small value of magnetization suggests that, with (Ca, Co) co-doping, the grown sample possesses a weak ferromagnetic state which is due to the predominant antiferromagnetic ordering of Cr^{3+} spins [29, 30].

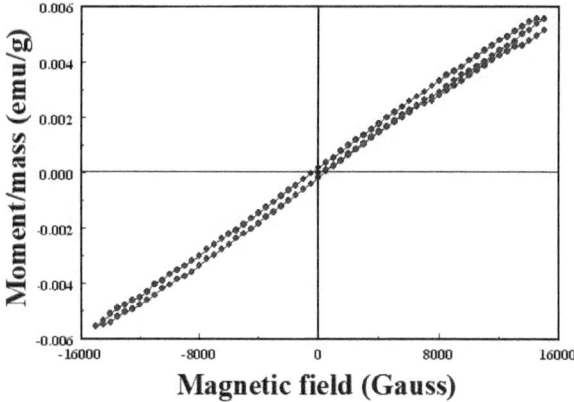

Figure 9 M-H curve for $(La_{0.8}Ca_{0.2})(Cr_{0.9}Co_{0.1})O_3$ obtained by vibrating sample magnetometry measurements.

Table 3 Magnetic parameters from VSM measurements for $(La_{0.8}Ca_{0.2})(Cr_{0.9}Co_{0.1})O_3$.

Parameters	Values
$M_s \times 10^{-3}$ (emu g^{-1})	5.55
H_c (G)	397.86
$M_r \times 10^{-3}$ (emu g^{-1})	179.43

4. CONCLUSION

The co-doped lanthanum chromite $(La_{0.8}Ca_{0.2})(Cr_{0.9}Co_{0.1})O_3$ was synthesized using the solid state reaction technique and characterized with the powder X-ray diffraction (PXRD) method to analyze the structural information. The spatial charge density distribution in the unit cell was investigated for the grown sample using PXRD through

MEM technique. Covalent nature of bonding was observed between the Cr and the O2 atoms and ionic nature of bonding was observed between the La and the O2 atoms. The UV-Visible absorption spectra give the energy band gap of the sample. The particle size of the prepared sample using SEM micrograph is found to be in the range of sub micrometers. VSM measurements show that the grown sample possesses weak ferromagnetism due to the predominant antiferromagnetic ordering due to Cr^{3+} spins.

REFERENCES

[1]   G. Setz Luiz Fernando, H. Sonia Regina, Mello-Castanho, Determining the Lanthanum Chromite Zeta Potential in Aqueous Media, Materials Science Forum 660-661 (2010) 1145-1150
      http://dx.doi.org/10.4028/www.scientific.net/MSF.660-661.1145

[2]   D. B. Meadowcroft, Some properties of strontium-doped lanthanum chromite, Brit. J. Appl. Phys. 2 (1969) 1225-1233
      http://dx.doi.org/10.1088/0022-3727/2/9/304

[3]   N. Russo, D. Fino, G. Sanacco, V. Speechia, Studies on the redox properties of chromite perovskite catalysts for soot combustion, Journal of catalysis 229 (2005) 459-469
      http://dx.doi.org/10.1016/j.jcat.2004.11.025

[4]   S. Ifrah, A. Kaddomi, P. Gelin, G. Bergeret, On the effect of La-Cr-O phase composition on diesel soot catalytic combustion, Catalysis Communications 8 (2007) 2257-2262
      http://dx.doi.org/10.1016/j.catcom.2007.04.039

[5]   S. A. Suvorov and A. P. Shevchick, A heating module equipped with Lanthanum Chromite - Based heaters, Refractories and Industrial ceramics 45 (2004) 196-200
      http://dx.doi.org/10.1023/B:REFR.0000036729.24986.e3

[6]   D. L. West, F. C. Montgomery, T. R. Armstrong, Use of La0.85Sr0.15CrO3 in high-temperature NOx sensing elements, Sensors and Actuators B 106 (2005) 758-765
      http://dx.doi.org/10.1016/j.snb.2004.09.028

[7]   [7]   W.L. David, F.C. Montgomery, T.R. Armstrong, "NO-selective" NOx sensing elements for combustion exhausts, Sensors and Actuators B 111-112 (2005) 84-90
      http://dx.doi.org/10.1016/j.snb.2005.06.043

[8]   Zhu Wei-zhong, Y. Mi, Perspectives on the metallic interconnects for solid oxide fuel cells, J. Zhejiang Univ Sci. 5(12) (2004) 1471-1503

http://dx.doi.org/10.1631/jzus.2004.1471

[9]   M. Suzuki, H. Sasaki, A. Kajimura, Oxide ionic conductivity of doped lanthanum chromite thin film interconnectors, Solid State Ionics 96 (1997) 83-88
       http://dx.doi.org/10.1016/S0167-2738(97)00007-6

[10]   K. Hilpert, D. Das, M. Miller, D. H. Peck and R. Wei, Chromium vapor species over solid oxide fuel cell interconnect materials and their potential for degradation processes, J. Electrochem. Soc. 143 No. 11 (1996) 3642-3647
       http://dx.doi.org/10.1149/1.1837264

[11]   H.M. Rietveld, A Profile Refinement Method for Nuclear and Magnetic structures, J. Appl. Crystallogr. 2 (1969) 65-71
       http://dx.doi.org/10.1107/S0021889869006558

[12]   V. Petricek, M. Dusek, L. Palatinus, Jana 2006, The Crystallographic Computing System, Institute of Physics, Prague, Czech Republic, (2006)

[13]   K.P. Ong, Peter Blaha, Ping Wu, Origin of the light green color and electronic ground state of LaCrO3, Physical review B 77 073102 (2008) 1-4.

[14]   K. Momma, F. Izumi, VESTA: a three-dimensional visualization system for electronic and structural analysis, J. Applied Crystallogr. 41 (2008) 653-658
       http://dx.doi.org/10.1107/S0021889808012016

[15]   T. Brajesh, A. Dixit, R. Naik, G. Lawes and M.S.Rama Chandra Rao, Magneto structural and magneto caloric properties of bulk LaCrO3 system, Materials Research Express 2 (2015) 1-13

[16]   R.D. Shannon, Revised effective ionic radii and systematic studies of interatomic distances in halides and chalcogenides. Acta Cryst. A32, (1976) 751-767.
       http://dx.doi.org/10.1107/S0567739476001551

[17]   D.M. Collins, Electron density images from imperfect data by iterative entropy maximization, Nature 49 (1982) 298.
       http://dx.doi.org/10.1038/298049a0

[18]   A. D. Ruben, I. Fugio, Superfast program PRIMA for the Maximum Entropy Method, Advanced Materials Laboratory, National Institute for Material Science, Ibaraki, Japan (2004), 3050044.

[19]   C.S. Montross, Elastic Modulus Versus Bond Length in Lanthanum Chromite Ceramics, Journal of the European Ceramic Society 18 (1997) 353-358.
       http://dx.doi.org/10.1016/S0955-2219(97)00143-X

[20]   S. Natsuko, F Helmer, C. Hauback, Structural, Magnetic and thermal properties of La1-tCatCrO3 J. of Solid state Chemistry 121 (1996) 202-213. http://dx.doi.org/10.1006/jssc.1996.0029

[21]   J. Tauc, R. Grigorvici, A. Vancu, Optical Properties and Electronic Structure of Amorphous Germanium, Physica Status Solidi 15, 627–637 (1966). http://dx.doi.org/10.1002/pssb.19660150224

[22]   S. Qing-Gong, S. Lingling, Z. Hui, W. Tong and K. Jianhai, The Structural Stabilities and Electronic Properties of Orthorhombic and Rhombohedral LaCrO3- A First principles Study, Advanced Materials Research 622-623 (2013) 734-738.

[23]   R.Saravanan, GRAIN software, Private Communication, (2008).

[24]   B.D. Culllity, S.R. Stock, Elements of X-ray Diffraction, third ed. Prentice Hall, New Jersy, 2001.

[25]   S.M. Khetre, H.V. Jadhav and S.R. Bamane, Synthesis and characterization of nanocrystalline LaCrO3 by combustion route, Rasayan J. Chem. 2 (2009) 174-178.

[26]   I. Masato, T. Hirotsugu, U. Kyota, E. Tadashi and S. Masahiko, Microwave synthesis of LaCrO3, Journal of Materials chemistry 8 (1998) 2765-2768. http://dx.doi.org/10.1039/a804139c

[27]   J.P. Gonjal, R. Schmidt, J.J. Romero, D.U. Amador and E. Moran, Microwave-Assisted Synthesis, Microstructure, and Physical Properties of Rare-Earth Chromites, Inorg. Chem. 2013, 52, 313−320. http://dx.doi.org/10.1021/ic302000j

[28]   G.A. Alvarez, X.L. Wang, G. Peleckis and S.X. Dou, J.G. Zhu and Z.W. Lin, Magnetotransport and magnetic properties of weak ferromagnetic semiconductors: Ca doped LaCrO3, Journal of applied physics 103 07B916 (2008) 1-3.

[29]   H. Terashita, J.C. Cezar, F.M. Ardito, L.F. Bufaical and E. Granado, Element-specific and bulk magnetism, electronic, and crystal structures of La0.70Ca0.30Mn1−xCrxO3, Physical Review B 85, 104401 (2012). http://dx.doi.org/10.1103/PhysRevB.85.104401

[30]   R. Shukla, J. Manjanna, A.K. Bera, S.M. Yusuf, and A.K. Tyagi, La1-xCexCrO3 (0.0 ≤x≤1.0): A New Series of Solid Solutions with Tunable Magnetic and Optical Properties, Inorg. Chem. 2009, 48, 11691–11696. http://dx.doi.org/10.1021/ic901735d

CHAPTER 5

Ferroelectric Charge Ordering in BaTi$_{0.9}$Zr$_{0.1}$O$_3$ Lead-Free Ceramics through Powder X-Ray Diffraction

J. Mangaiyarkkarasi[1*], R. Saravanan[2]

[1]PG and Research Department of Physics, NMSSVN College, Nagamalai, Madurai-625 019, Tamil Nadu, India

[2]Research Centre and PG Department of Physics, The Madura College, Madurai-625 011, Tamil Nadu, India

*mangai.jp@gmail.com

Abstract

In this work, we report the structural, ferroelectric charge ordering and optical properties of lead-free BaTi$_{0.9}$Zr$_{0.1}$O$_3$ ceramics synthesized by the high temperature solid state reaction method. The grown sample was characterized using powder X-ray diffraction, Rietveld refinement, UV-visible spectroscopy and scanning electron microscopy. The tetragonal structure of the sample was confirmed through the X-ray data. Two dimensional MEM charge density maps and one dimensional profiles revealed the enhanced covalent character between titanium and oxygen ions and the ionic nature between barium and oxygen ions. The energy band gap is evaluated as 2.87eV from UV-vis analysis. The surface morphology is analyzed with the SEM micrograph.

Keywords

Ceramics, X-Ray Diffraction, Charge Density, Rietveld Refinement, Maximum Entropy Method

Contents

1.   INTRODUCTION

Complex perovskite structures with formula ABO_3 are widely used for computer memories, pyroelectric detectors, multilayer capacitors and other electronic devices [1,2]. Particularly $PbTiO_3$ ceramics have dominated over several decades in the technological field. Due to the toxic and hazardous behavior of Pb-based materials, recent research has been intensified for lead-free perovskites [3]. Among the available lead-free materials zirconium doped $BaTiO_3$ (BZT) shows promising piezo electric and electrostrictive properties [4]. Substitution of Ti^{4+} with Zr^{4+} ions significantly enhances the electrical and structural properties of $BaTiO_3$ to a larger extend [5]. Moreover Zr^{4+} ion is more stable than Ti^{4+}, so the substitution of Ti by Zr would depress the conduction thereby decreasing the leakage current in the $BaTiO_3$ system [6]. BZT ceramics possess two main advantages over Sr doped $BaTiO_3$ such as low dielectric loss and a high dielectric constant [7]. Aiyun Liu et al. [8] investigated the infrared and optical properties of BZT ceramics, which are essential to optimize the design and understanding the devices related to pyroelectric and electric-optical applications. Even though low temperature chemical methods have shown potential in the formation of pure BZT, they are not able to enhance the electrical properties for the desired applications. But the high temperature solid state reaction with high energy ballmilling promotes good grain growth and shows significant improved properties [9, 10]. Zhao Xin-Yin et.al [11] has reported the structural, electronic and optical properties of BZT through first principles study. In our study, we have analyzed the ferroelectric charge ordering and optical properties of $BaTi_{0.9}Zr_{0.1}O_3$ ceramic material. Elbasset et.al [12] observed that, increasing Zr content in the $BaTiO_3$ based compositions caused a drop in the Curie temperature (Tc) and increase in permittivity. The dielectric loss is lowered down to 90% when Zr content is added by 15% , indicating the ceramic with Zr=0.15 is well suitable for tunable capacitor applications. The experimental electron density studies for the Zr doped $BaTiO_3$ are hardly available in current literature, so we have concentrated more on the bonding nature between the atoms, bond length and electron density distribution between the atoms inside the unit cell. This can be achieved by adopting the maximum entropy method (MEM) [13] which was successfully introduced by Collins in 1982.

2. SYNTHESIS AND CHARGE DENSITY STUDIES

2.1 MATERIAL SYNTHESIS

Ceramic system $BaTi_{0.9}Zr_{0.1}O_3$ was prepared using the conventional high temperature solid state reaction technique using high purity starting materials $BaCO_3$ (Alfa aesar 99.997%), ZrO_2 (Alfa aesar 99.99%) and TiO_2 (Alfa aesar 99.99%). Stoichiometrically weighed powders were thoroughly ground using agate mortar and pestle and calcined at 1200° C for 2 h in an alumina crucible using a tubular furnace. Then the calcined powder was ground well using a ballmill at 200 rpm for 5 h and compressed into dense pellets. These compacts were sintered at $1450\ ^{\circ}C$ with a soaking time of 10 h in air with a heating rate of 5 $^{\circ}C/min$ and then they were slowly cooled at a normal cooling rate of the furnace. Finally the samples were taken out and ground well as smooth powder.

2.2 CHARACTERIZATION TECHNIQUES

The grown sample was characterized by powder XRD intensity data using a X-ray diffractometer (Bruker AXS D8 advance) with CuKα, monochromatic incident beam (λ=1.54056Å), obtained in the 2θ range of 10°-120° with the step size of 0.02 and band gap estimated from the UV-vis data from UV-vis spectrometer (Cary 5000, Varian, Germany) in the range of 200 nm-2000 nm collected at Sophisticated Analytical Instrument Facility (SAIF), Cochin University, Cochin. SEM images were recorded using a scanning electron microscope (Carl Zeiss Evo 18) corresponding to various magnifications (×5000, ×10000, ×25000) analyzing the surface morphology, obtained at International Research Centre, Kalasalingam University, Krishnan Coil.

2.3 X-RAY DIFFRACTION

The raw XRD powder pattern of $BaTi_{0.9}Zr_{0.1}O_3$ is shown in fig 1. The prepared BZT ceramics presents the tetragonal structure with space group P4mm (No.99) in agreement with the Joint Committee on Powder Diffraction Standards (JCPDS #05-0626).

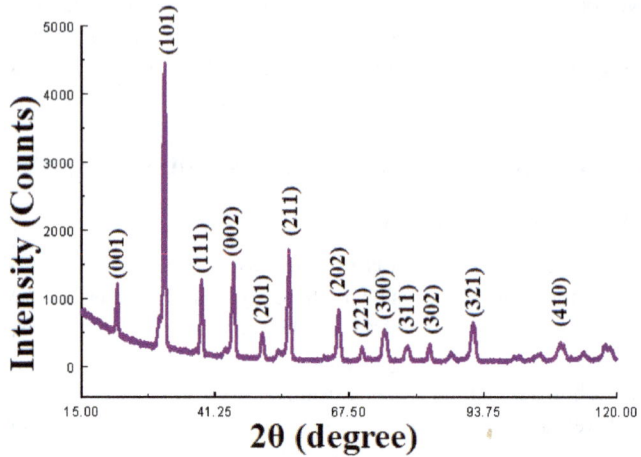

Fig. 1 Experimental X-ray diffraction profile of BaTi$_{0.9}$Zr$_{0.1}$O$_3$

Lattice parameters, cell volume and structure factors are calculated through Rietveld refinement using the software package JANA 2006 [14]. The refinement was performed by considering one formula unit per unit cell with a P4mm space group and the atomic coordinates (x,y,z) were initialized as (0,0,0) for barium, (0.5, 0.5, 0.512) for titanium, (0.5,0.5,0.023) for oxygen1 and (0.5,0,0.486) for oxygen2 [15].

Fig. 2 depicts the refinement plot of the ceramic sample grown at 1450 °C for 10 h, in which the observed XRD profiles are indicated by dots and the calculated profiles are represented by the continuous lines. The difference between the observed profile and the calculated profile is shown at the bottom of the figure and the Bragg peak positions are intimated by the small vertical lines below the fitted profile. Refinement results are given in Table 1.

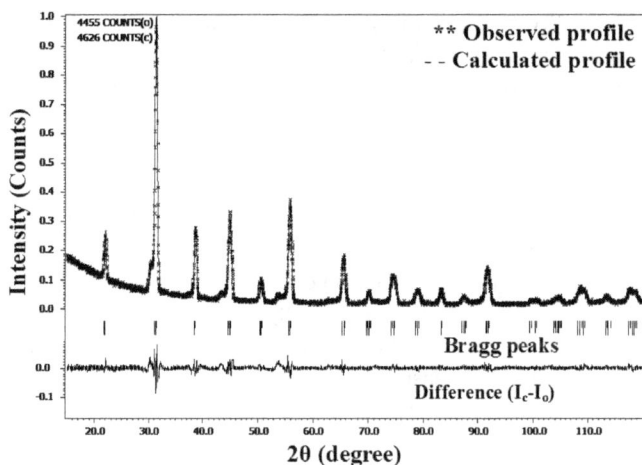

Fig. 2. Rietveld refinement plot of $BaTi_{0.9}Zr_{0.1}O_3$

Table .1. Structural parameters of $BaTi_{0.9}Zr_{0.1}O_3$ from Rietveld refinement

Parameters	Values
a(Å)	3.999(21)
c(Å)	4.027(18)
$\alpha = \beta = \gamma$	90
Cell Volume (Å^3)	64.41(15)
Density (gm/cc)	6.121(24)
R_{obs} (%)	2.14
wR_{obs} (%)	2.70
R_p (%)	7.72
wR_p (%)	2.70
GOF	1.75

The refined lattice parameter values (a=b=3.999 Å, c=4.027 Å) and the volume of the unit cell (V=64.41 Å^3) are in good agreement with those reported in the literature for BZT [16].

Average grain size was calculated from the peak broadening using the most intense five peaks at 2θ =22.09°, 31.35°, 38.62°, 44.90° and 55.88° by employing the Debye-Scherrer equation (t = 0.9λ / βcosθ, where t is grain size, λ is wavelength of X-ray, β is the full width at half maximum and θ is the Bragg angle) through the Grain software [17]. The resultant average grain size is found to be 25 nm which is in agreement with the reported values [18].

2.4   CHARGE DENSITY DISTRIBUTION

To understand the accurate electronic structure and the nature of chemical bonds precisely, the charge density distribution analysis is utmost important. In this regard the maximum entropy method (MEM) [13] is an exact versatile tool to elucidate the electron density. Charge density distribution is successfully calculated using the software package PRIMA [19] by using the structure factors derived from the Rietveld analysis. The 3-dimensional unit cell and 2-dimensional electron density contour maps are plotted with the help of the visualization software VESTA [20]. Position of the Ba, the Ti and the O atoms along with the regions of electron density around them is clearly visualized in the 3-dimensional unit cell shown in Fig. 3.

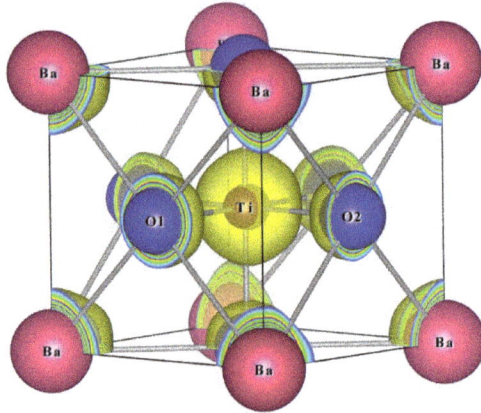

Fig. 3. 3-dimensional electron density distribution of $BaTi_{0.9}Zr_{0.1}O_3$

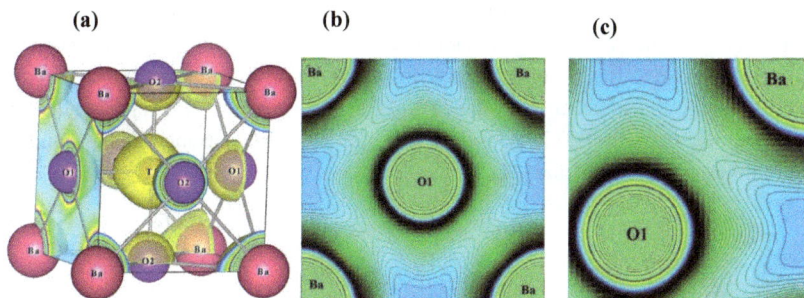

Fig. 4(a) 3D unit cell of BaTi$_{0.9}$Zr$_{0.1}$O$_3$with (001) plane shaded (b) 2D electron density map on (001) plane (c) Enlarged view of Ba-O1 bond

Fig 4(a) shows the unit cell of BaTi$_{0.9}$Zr$_{0.1}$O$_3$ with (001) plane shaded and the 2D electron density map corresponding to the plane (001). Fig 4(c) depicts the enlarged view of Ba-O1 bond on the same plane. Contour lines around the Ba and O1 atoms are not much linked with each other resulting from a lesser electron density between them.

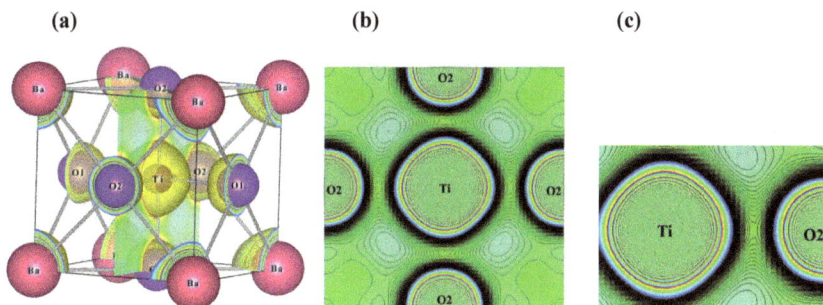

Fig. 5(a) 3D unit cell of BaTi$_{0.9}$Zr$_{0.1}$O$_3$with (002) plane shaded (b) 2D electron density map on (002) plane (c) Enlarged view of Ti-O2 bond

Fig. 5(a) shows the 3D unit cell indicating the (002) plane and 5(b) represents the corresponding 2-dimensional electron density map. Fig. 5(c) represents the enlarged view of the Ti-O2 bond. The density of contour lines around the Ti and the O2 atoms exhibits maximum charge linkage between the Ti and the O2 atoms indicating the covalent character of Ti-O2 bond. The same behavior is observed in the (101) plane also shown in fig. 6(b) which is due to the strong hybridization of Ti 3d orbitals and O 2p orbitals and their contribution to the states on the valence and conduction bands. It is reported that the origin of ferroelectricity in $BaTiO_3$ strongly depends on the occurrence of the bonding between titanium and oxygen ions [21].

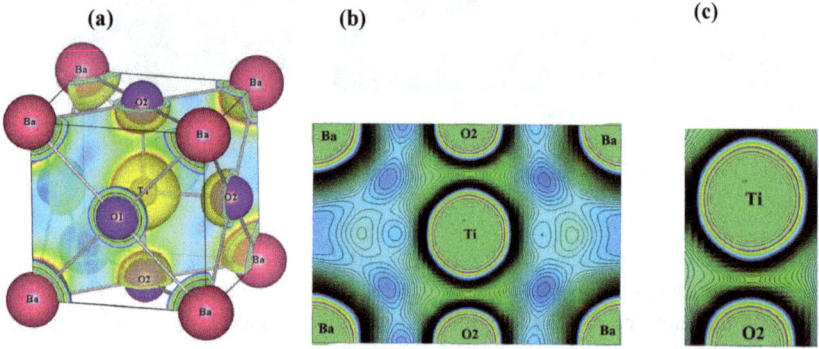

Fig. 6(a) 3D unit cell of $BaTi_{0.9}Zr_{0.1}O_3$ with (101) plane shaded (b) 2D electron density map on (101) plane (c) Enlarged view of Ti-O2 bond

Fig. 6(a) shows the 3D unit cell with (101) plane shaded and 6(b) represents the 2-dimensional electron density map. An enlarged view of the Ti-O2 bond is shown in fig. 6(b). Fig. 7(a) & 7(b) represent the one dimensional electron density profiles of (Ba-O1) and (Ba-O2) bonds. Fig. 8(a) & 8(b) represent the one dimensional electron density profiles of (Ti-O1) and (Ti-O2) bonds. Table 2 gives the bond lengths and mid bond density values between the Ba-O1, Ba-O2, Ti-O1 and Ti-O2 bonds.

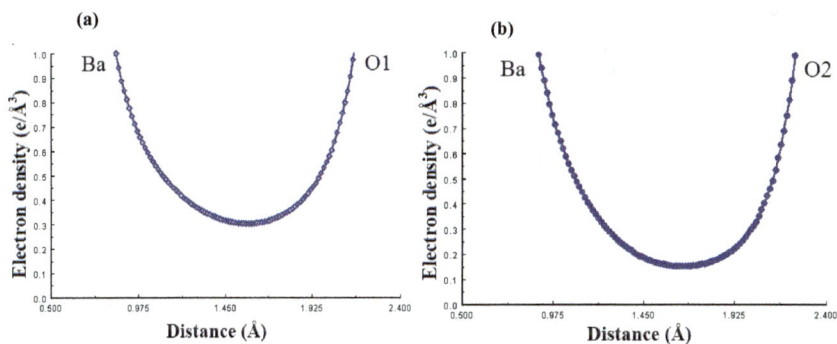

Fig. 7(a) one dimensional electron density profile for (Ba-O1) bond (b) one dimensional electron density profile for (Ba-O2) bond

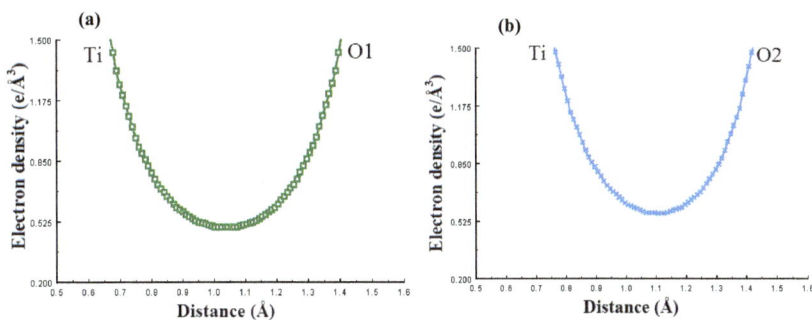

Fig. 8(a) one dimensional electron density profile for (Ti-O1) bond (b) one dimensional electron density profile for (Ti-O2) bond

The mid bond density between Ba and O1 atoms is 0.2007 e/$Å^3$, confirming the ionic nature between barium and oxygen and 0.6174 e/$Å^3$ between Ti and O2 atoms, which denotes the covalent character between titanium and oxygen.

Table .2. One dimensional electron density values

Bond	Bond length (Å)	Mid bond density (e/Å³)
(Ba-O1)	2.837	0.2007
(Ba-O2)	2.827	0.2260
(Ti-O1)	2.013	0.5948
(Ti-O2)	1.999	0.6174

2.5   OPTICAL STUDIES

From the UV-visible absorption spectral analysis, the band gap of the grown sample can be determined.

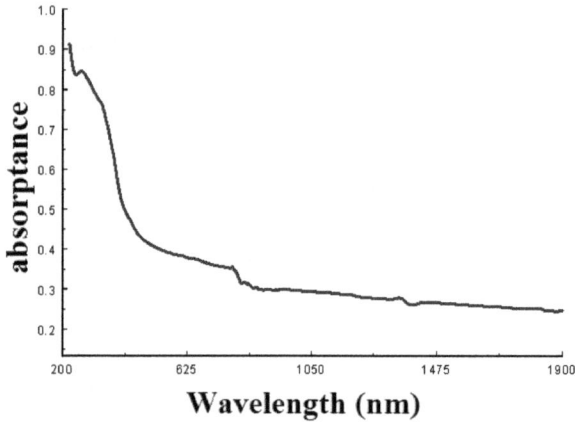

Fig. 9. Absorption spectrum of BaTi$_{0.9}$Zr$_{0.1}$O$_3$

Fig. 9 shows the optical absorption spectrum of BaTi$_{0.9}$Zr$_{0.1}$O$_3$ which shows the absorption peak at 264nm. The band gap is estimated by modifying the absorption spectrum using the relation proposed by Wood and Tauc [22] which is given as $\alpha h v = B (h v - E_g)^n$, where B is the material constant, α is absorption coefficient, hv is photon energy, E_g is the energy band gap, n=1/2 corresponds to direct band gap materials and n=2 for indirect band gap materials. Fig. 10 shows the plot of energy verses $(\alpha h v)2$. From this plot, the extrapolation of the tangent of straight line portion of the curve to X-axis, gives the value of energy gap Eg.

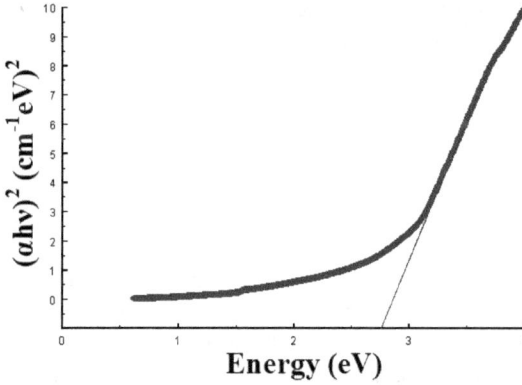

Fig. 10 The plot of energy Vs $(\alpha h v)^2$ for $BaTi_{0.9}Zr_{0.1}O_3$

The energy gap value of $BaTi_{0.9}Zr_{0.1}O_3$ estimated as 2.87 eV is in good agreement with the reported values for Zr doped $BaTiO_3$ systems [23].

2.6 SEM ANALYSIS

Surface morphology and the microstructure of the prepared ceramics were characterized using the scanning electron microcopy (SEM).

Fig. 11. SEM image of $BaTi_{0.9}Zr_{0.1}O_3$ at the magnification level of ×5000.

Fig. 11 shows the SEM micrograph of $BaTi_{0.9}Zr_{0.1}O_3$ corresponding to the magnification ×5000. It can be observed that the prepared sample is composed of small particles possessing non uniform microstructures. The average particle size is determined as 2 μm [24].

CONCLUSION

In summary, $BaTi_{0.9}Zr_{0.1}O_3$ was synthesized using the solid state reaction method at $1450^{\circ}C$ for 10 h. XRD profiles and Rietveld refinement confirmed that the grown ceramic crystallized in a tetragonal structure with lattice constants (a=b=3.999 Å, c=4.027 Å) and cell volume 64.41 $Å^3$. MEM charge density analysis exhibits the enhancement of a covalent bond character between the titanium and the oxygen ions and an ionic character between the barium and the oxygen ions. Calculated average grain size is 25 nm. The energy band gap is estimated as 2.87 eV using the UV-vis analysis. The SEM micrograph reveals the formation of non uniform particles.

REFERENCES

[1]   Fangyi Rao, Miyoung Kim, A. J. Freeman, Shaoping Tang and Mark Anthony, Structural and electronic properties of transition-metal/$BaTiO_3$(001) interfaces, Physical Review B, 55, (1997) 953-960.
       http://dx.doi.org/10.1103/PhysRevB.55.13953

[2]   Zhang Guang zu, Yi Jinqiao, Jiang Shenglin,Yu Yan, He Jungang, Liu Sisi, Zhu Dingyang, Zhang Ling, Role of internal stress on dielectric and dc bias field-induced pyroelectric properties of $Ba_{0.68}Sr_{0.32}TiO_3$-$(Ba_{0.68}Sr_{0.32})_2TiO_4$ for uncooled infrared detectors, http://www.paper.edu.cn.

[3]   J.C.Sczancoski, L.S. Cavalcante, T. Badapanda, S.K. Rout, S. Panigrahi,V.R. Mastelaro, J.A. Varela, M. Siu Li, E. Longo, Structure and optical properties [$Ba_{1-x}Y_{2x/3}$]($Zr_{0.25}Ti_{0.75}$)O_3 powders, Solid State Sciences 12 (2010) 1160-1167.
       http://dx.doi.org/10.1016/j.solidstatesciences.2010.04.002

[4]   Sandeep Mahajan, O.P. Thakur, D.K. Bhattacharya, K. Sreenivas, A comparative study of $Ba_{0.95}Ca_{0.05}Zr_{0.25}Ti_{0.75}O_3$ relaxor ceramics prepared by conventional and microwave sintering techniques, Materials Chemistry and Physics 112 (2008) 858–862.
       http://dx.doi.org/10.1016/j.matchemphys.2008.06.054

[5]   P. Sateesh, J. Omprakash, G. S. Kumar and G. Prasad, Studies of phase transition and impedance behavior of Ba(Zr,Ti)O_3 ceramics, Journal of Advanced Dielectrics 5, (2015) 1-13.

http://dx.doi.org/10.1142/S2010135X15500022

[6]   S. Sarangi , T. Badapanda , B. Behera ,S. Anwar, Frequency and temperature dependence dielectric behavior of barium zirconate titanate nanocrystalline powder obtained by mechanochemical synthesis, J Mater Sci: Mater Electron.

[7]   S. Parida, S.K. Rout, L.S. Cavalcante, A.Z. Simões, P.K. Barhai, N.C. Batista, E. Longo, M. Siu Li, S.K. Sharma, Structural investigation and improvement of photoluminescence properties in $Ba(Zr_xTi_{1-x})O_3$ powders synthesized by the solid state reaction method, Materials Chemistry and Physics 142 (2013) 70-76.
http://dx.doi.org/10.1016/j.matchemphys.2013.06.041

[8]   Aiyun Liu, Jianqiang Xue, Xiangjian Meng, Jinglan Sun, Zhiming Huang, Junhao Chu, Infrared optical properties of $Ba(Zr_{0.20}Ti_{0.80})O_3$ and $Ba(Zr_{0.30}Ti_{0.70})O_3$ thin films prepared by sol–gel method, Applied Surface Science 254 (2008) 5660–5663.
http://dx.doi.org/10.1016/j.apsusc.2008.03.178

[9]   S.K. Rout, L.S.Cavalcante, J.C.Sczancoski, T.Badapanda, S.Panigrahi, M.SiuLi, E.Longo, Photo luminescence property of $Ba(Zr_{0.25}Ti_{0.75})O_3$ powders prepared by solid state reaction and polymeric precursor method, Physica B 404 (2009) 3341–3347.
http://dx.doi.org/10.1016/j.physb.2009.05.014

[10]   Wei Cai, Chunlin Fu, Jiacheng Gao, Huaqiang Chen, Effects of grain size on domain structure and ferroelectric properties of barium zirconate titanate ceramics, Journal of Alloys and Compounds 480 (2009) 870–873.
http://dx.doi.org/10.1016/j.jallcom.2009.02.049

[11]   Zhao Xin-Yin, Wang Yue-Hua, Zhang Min, Zhao Na, Gong Sai, Chen Qiong, First-Principles Calculations of the Structural, Electronic and Optical Properties of $BaZr_xTi_{1-x}O_3$ (x = 0, 0.25, 0.5, 0.75), chin. phys. lett. 28, (2011) 1-5.

[12]   A.Elbasset, E.Abdi, T.Lamcharfi, S. Sayouri,M.Abarkan, N.S.Echatoui and M.Aillerie, Influence of Zr on structure and dielectric behaviour of $BaTiO_3$ ceramics. Indian J.of.Science and Technology 8(2015)
http://dx.doi.org/10.17485/ijst/2015/v8i13/56574

[13]   D.M. Collins. Electron density images from imperfect data by iterative entropy maximization. Nature., 298 (1982) 49-51.
http://dx.doi.org/10.1038/298049a0

[14] V. Petricek, M. Dusek, L. Palatinus, The Crystallographic Computing System JANA 2006. (Institute of Physics, Academy of Sciences of the Czech Republic, Praha, (2000).

[15] R.W.G. Wyckoff, Crystal structures. 1 Inter-space publishers, London, 1963.

[16] N. Sawangwan, J. Barrel, K. Mackenzie, T. Tunkasiri, The effect of Zr content on electrical properties of $Ba(Ti_{1-x}Zr_x)O_3$ ceramics, Appl. Phys. A 90, (2008) 723–727.
http://dx.doi.org/10.1007/s00339-007-4342-9

[17] R. Saravanan, Grain software (Personal communication) (2008).

[18] Lavinia Petronela Curecheriu, Raluca Frunza and Adelina Ianculescu, Dielectric properties of the $BaTi_{0.85}Zr_{0.15}O_3$ ceramics prepared by different techniques, Processing and Application of Ceramics 2 (2008) 81–88.
http://dx.doi.org/10.2298/PAC0802081C

[19] A.D. Ruben, F. Izumi, Super-fast Program PRIMA for the Maximum-Entropy Method, Advanced Materials Laboratory (National institute for materials science, Tsukuba, Ibaraki, 2004).

[20] K. Momma, F. Izumi, VESTA: a three-dimensional visualization system for electronic and structural analysis Journal of Applied Crystallography., 41, 2008, 653.
http://dx.doi.org/10.1107/S0021889808012016

[21] Colin L. Freeman ,James A. Dawson , Hung-Ru Chen , Liubin Ben , John H. Harding Finlay D. Morrison , Derek C. Sinclair , and Anthony R. West Energetics of Donor-Doping, Metal Vacancies, and Oxygen-Loss in A-Site Rare-Earth-Doped $BaTiO_3$ www.material views.com.

[22] J. Tauc, R. Grigorvici,Y. Yanca, Optical properties and electronic structure of amorphous germanium. Phys. Status Solidi. 1966, 15, 627–637.
http://dx.doi.org/10.1002/pssb.19660150224

[23] T Badapanda, S K Rout, L S Cavalcante, J C Sczancoski, S Panigrahi, E Longo and M Siu Li, Optical and dielectric relaxor behavior of $Ba(Zr_{0.25}Ti_{0.75})O_3$ ceramic explained by means of distorted clusters, J. Phys. D: Appl. Phys. 42 (2009) 175414.
http://dx.doi.org/10.1088/0022-3727/42/17/175414

[24] C. E. Ciomaga, R. Calderone, M. T. Buscaglia, M. Viviani, V. Buscaglia, L. Mitoseriu, A. Stancu, P. Nanni, Relaxor properties of $Ba(Zr,Ti)O_3$ ceramics, Journal of Optoelectronics and Advanced Materials 8, (2006) 944 – 948.

CHAPTER 6

Electronic Bonding Analysis on Dilute Doping of Iron in Nickel Oxide Nano Crystals

K. S. Arjun[1], S. Saravanakumar[2*], M. Prema Rani[1]

[1]Research Centre and PG Department of Physics, The Madura College, Madurai-625011, Tamil Nadu, India

[2]Department of Physics, Kalasalingam University, Krishnankoil, Viruthunagar - 626 126

*Email: premaakumar@yahoo.com

Abstract

The electronic bonding of $Ni_{1-x}Fe_xO$ (x=0.00, 0.01, 0.02) nanocrystalline powders was analyzed using MEM (Maximum Entropy Method). All the samples were prepared using the chemical co-precipitation method. X-ray diffraction (XRD), UV absorption spectroscopy and Vibration Sample Magnetometer (VSM) measurements were performed to study the crystal structure, optical bang gap and magnetic properties of the prepared samples. The X-ray data were refined using the Rietveld refinement. Using one, two and three dimensional MEM maps, the bonding within the atoms was evaluated.

Keywords

X-Ray Diffraction, Rietveld Refinement, Band Gap, Maximum Entropy Method.

Contents

1.   INTRODUCTION

The application of DMS materials (Diluted Magnetic Semiconductors) in spintronic devices has attracted considerable attention in research in recent years [1]. DMS materials have a wide novel application combining magnetic, electronic and optical functionalities. Semiconducting materials are characterized by the random substitution of a fraction of the original atoms by magnetic atoms. The materials are semi magnetic semiconductors (SMSC) or diluted magnetic semiconductors (DMS). Several of the properties of these materials may be tuned by changing the concentration of magnetic ions. Experiments have been carried out to study the properties of oxide-based DMS (e.g., ZnO, TiO_2 etc.) with various transition metal (TM) ions doped [2]. Stoichiometric NiO which is antiferromagnetic at room temperature exhibits ferromagnetism when converted into p-type by doping Fe [3].

In this research work, $Ni_{1-x}Fe_xO$ (x=0.00, 0.01, 0.02) nanoparticles are synthesized with the simple co-precipitation method and their charge density and bonding are determined by a statistical approach, the Maximum Entropy Method [4].

2.   SAMPLE PREPARATION

$Ni_{1-x}Fe_xO$ samples with x=0.00, 0.01 and 0.02 were prepared with the chemical co-precipitation method. Stoichiometric amounts of Ni $(NO_3)_2.6H_2O$ and Fe $(NO_3)_2.6H_2O$ were dissolved in de-ionised water. Aqueous NH_4CO_3 was added with magnetic stirring to acquire pH = 8. The resultant gels were centrifuged and dried. The samples were then calcined at 773K for 4 hours to obtain Fe doped NiO nanopowders.

3.   RESULTS AND DISCUSSION

3.1   POWDER X-RAY STUDIES

The powder X-ray data set was collected in the 2θ range from 10° to 120° with step size of 0.05° at SAIF, Cochin, with a monochromatic incident beam of the wavelength

1.54056 Å offering pure CuKα radiations. The obtained data was compared with JCPDS and the crystal structure was confirmed as cubic, with space group Fm$\bar{3}$m (225) with lattice constant 4.177 Å from the data (JCPDS No. 471049). The powder X-ray profiles for the prepared samples are shown in figure 1.

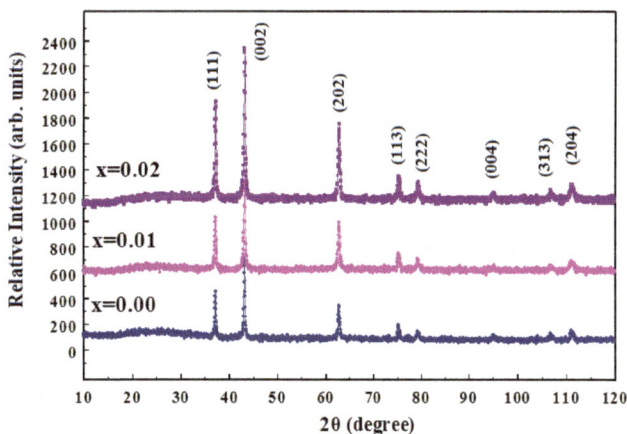

Figure 1: Powder XRD profiles for $Ni_{1-x}Fe_xO$.

3.2 RIETVELD ANALYSIS

The Rietveld refinement [5] is the standard tool which was devised by Hugo Rietveld [5] for use in the characterization of crystalline materials. In the present work, the cell parameters and other structural parameters were refined using the software, JANA 2006 [6]. The refined powder XRD profile for NiO is shown in figure 2 (a)-(c). The cell and other structural parameters are given in table 1.

3.3 CHARGE DENSITY USING THE MAXIMUM ENTROPY METHOD

The electron density distribution is analyzed using MEM. The bonding nature and the distribution of electrons in the bonding region can be clearly visualized using this technique [7]. For the numerical MEM computations on Fe doped NiO, the software package PRIMA [7, 8] was used. For the 2D and 3D representation of the electron densities, the program VESTA package was used [8]. The MEM refinements were carried out by dividing the unit cell into 48x48x48 pixels. The electron density at each pixel was fixed uniformly as F_{000}/a_0^3 e/Å3, where F_{000} is the total number of electrons in the unit cell and a_0 is the cell parameter. The MEM parameters are given in table 2. The 3D electron densities are shown in figures 3(a)-(c) for NiO, $Ni_{0.99}Fe_{0.01}O$, $Ni_{0.98}Fe_{0.02}O$. The 2D electron densities are shown in figures 4 (a)-(c) for NiO, $Ni_{0.99}Fe_{0.01}O$, $Ni_{0.98}Fe_{0.02}O$ on the (100) plane.

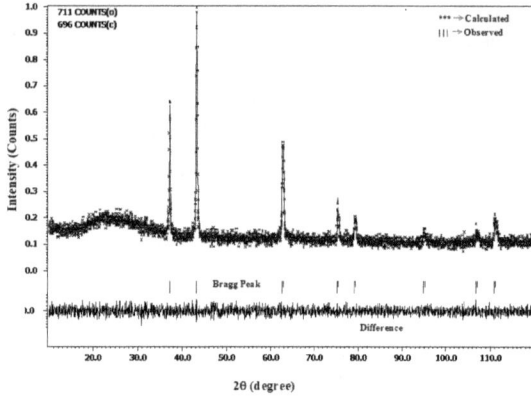

Figure 2(a): Rietveld refinement profile for NiO.

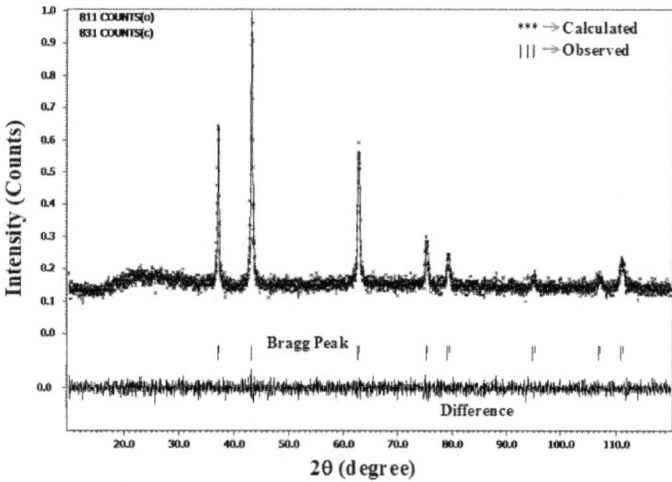

Figure 2 (b): Rietveld refinement profile for $Ni_{0.99}Fe_{0.01}O$.

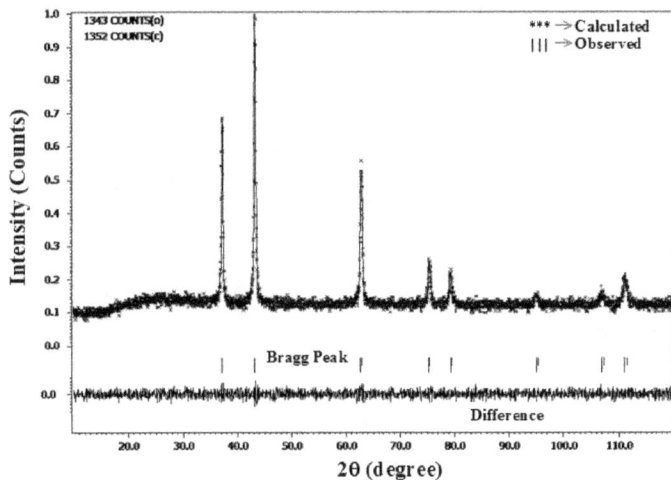

Figure 2(c): Rietveld refinement profile for $Ni_{0.98}Fe_{0.02}O$.

Table 1. Structural parameters of NiO, and Fe doped NiO.

Parameters	NiO	$Ni_{0.99}Fe_{0.01}O$	$Ni_{0.98}Fe_{0.02}O$
a=b=c (Å)	4.180784	4.179770	4.168728
B* - Ni	1.068206	0.992328	1.007899
B* - O	0.618814	0.964372	0.875182
R_{obs}	1.68	1.18	1.19
$_wR_{obs}$	1.33	0.99	0.97
R_p	6.97	5.82	5.05
$_wR_p$	8.89	7.50	6.44

B* - Debye Waller factor

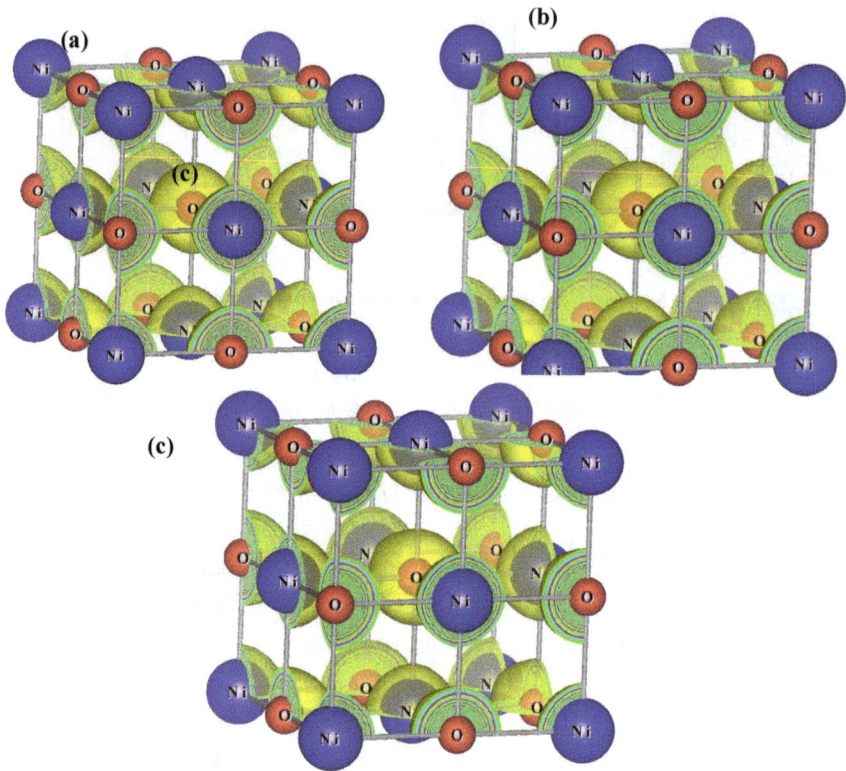

Figure 3 (a)-(c): Three dimensional electron density distribution in the unit cell (isosurface level = 1 e/$\mathring{A}^3$) for for $Ni_{1-x}Fe_xO$.

(a) **(b)** **(c)**

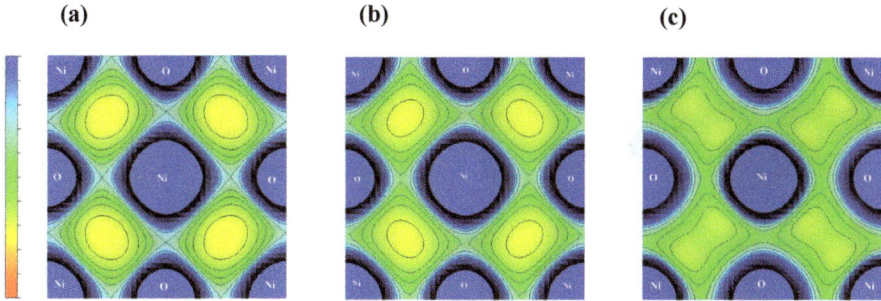

Figure 4(a)-(c): Two dimensional electron density distribution of $Ni_{1-x}Fe_xO$ along (100) plane in the contour range of 0 to 1.0 $e/Å^3$ with contour interval of 0.05 $e/Å$.

Table 2. MEM Parameters.

Parameters	NiO	$Ni_{0.99}Fe_{0.01}O$	$Ni_{0.98}Fe_{0.02}O$
Number of pixels in the unit cell	48X48X 48	48X48X48	48X48X 48
Number of Cycles	260	219	213
Lagrange Parameter (λ)	0.006295	0.006090	0.006060
R_{MEM} %	0.018575	0.017326	0.013781
wR_{MEM} %	0.019990	0.019635	0.015445

The 2D electron densities are shown in figure 5 (a)-(c) for NiO, $Ni_{0.99}Fe_{0.01}O$, $Ni_{0.98}Fe_{0.02}O$ on the (110) plane. The interaction of Nickel and Oxygen atoms are clearly visible. The covalent bonding between Nickel and Oxygen atoms is clearly visualized in the figure 5(a) and 5(c). High thermal vibration is observed for the oxygen atom in $Ni_{0.99}Fe_{0.01}O$ compared to NiO and $Ni_{0.98}Fe_{0.02}O$. The enlarged valence region $Ni_{0.99}Fe_{0.01}O$ is also due to this reason.

(a)   **(b)**

(c)

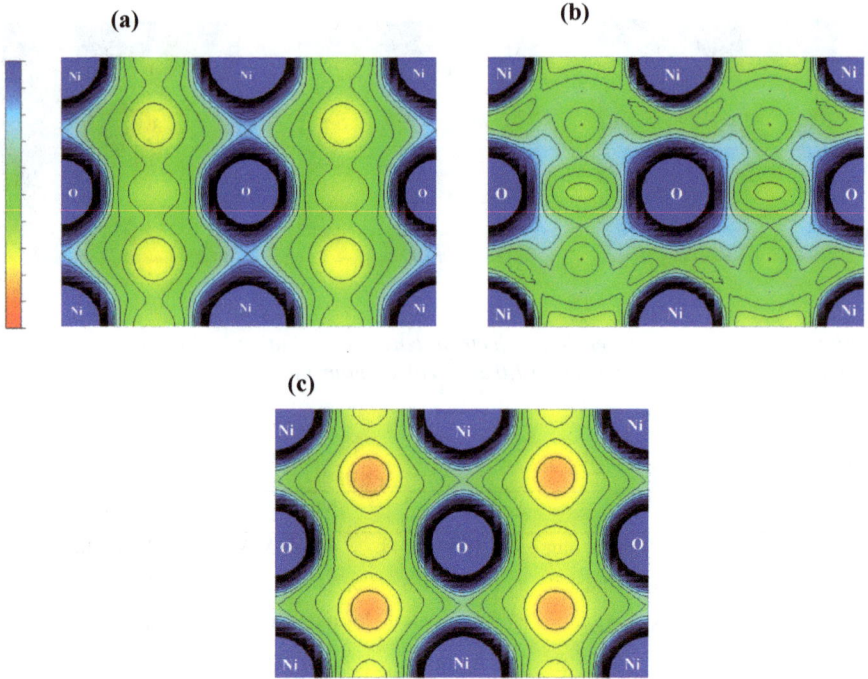

Figure 5 (a)-(c): Two dimensional electron density distribution of $Ni_{1-x}Fe_xO$ along (110) plane in the contour range of 0 to 1.0 $e/Å^3$ with contour interval of 0.05 $e/Å$.

One dimensional charge density for Ni and O atoms along (110) plane is shown in figure 6. The numerical data of electron density is given in table 3. The quantitative values for the charge density between nickel and oxygen atoms shown in table 3 confirm the covalent bonding.

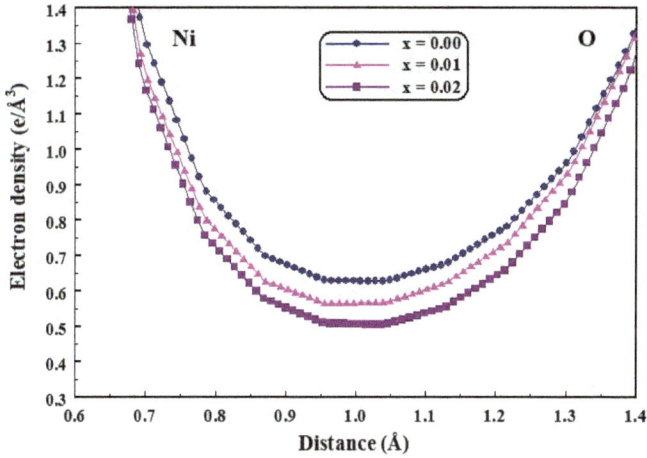

Figure 6: One dimensional charge density for Ni and O atoms along (110) plane.

Table 3. One dimensional electron density for Ni and O atoms.

$Ni_{1-x}Fe_xO$	Distance (Å)	Electron density $(e/Å^3)$
x = 0.00	1.0390	0.6269
x = 0.01	0.9662	0.5629
x = 0.02	1.0366	0.5043

3.4   UV-VISIBLE ANALYSIS

The band gap analysis is extracted from the UV data collected from SAIF, Cochin in the wavelength range of 200 to 800 nm. From the UV plot, the energy band gap is measured with the help of the Tauc relation, $\alpha h v = A(h v - E_g)^n$, where, α is the absorption coefficient, h is the Plank's constant and A is the optical constant v is the frequency and E_g is the energy band gap. The energy band spectrum is shown in figure 7 and the band gap values are tabulated in table 4. The band gap decreases for $Ni_{0.99}Fe_{0.01}O$ and it is higher for the other two samples. The enhanced electron density observed in figure 5b complements the result of the UV analysis.

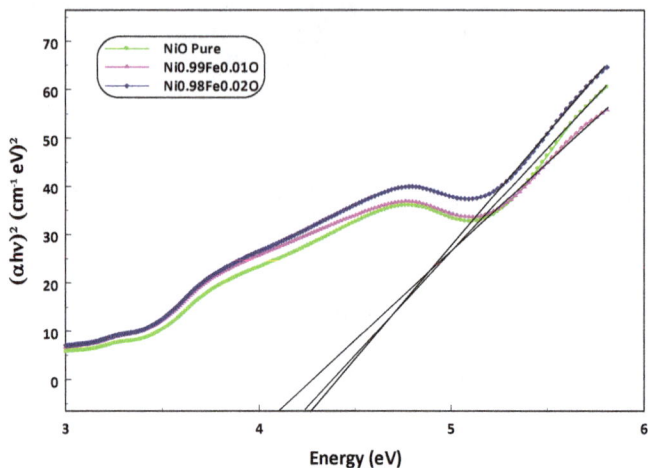

Figure 7: UV-Visible spectra for $Ni_{1-x}Fe_xO$.

Table 4. Band gap energy for $Ni_{1-x}Fe_xO$.

Samples	Band gap (eV)
NiO	4.2362
$Ni_{0.99}Fe_{0.01}O$	4.1036
$Ni_{0.98}Fe_{0.02}O$	4.2651

3.5   MAGNETIC PROPERTIES

The magnetic properties of $Ni_{1-x}Fe_xO$ were characterized by a vibrating sample magnetometer and the measurements were performed at IIT, Chennai. The magnetic properties strongly depended on the sample preparation and annealing temperature. Magnetic hysteresis curves were studied for the grown powder materials using a vibrating sample magnetometer (VSM) through measuring the magnetization M (emu/g) as a function of magnetic field H (G). Figure 8 shows the M-H curves taken at room temperature for the $Ni_{1-x}Fe_xO$ powder samples. Table 5 shows the parameter of $Ni_{1-x}Fe_xO$ from VSM measurements. The addition of Fe in NiO lattice has enhanced the magnetic property of NiO. The hysteresis pattern of VSM clearly shows the ferromagnetic nature of $Ni_{0.98}Fe_{0.02}O$. Considerable enhancement of magnetic effect for the Fe doped samples is observed in table 5.

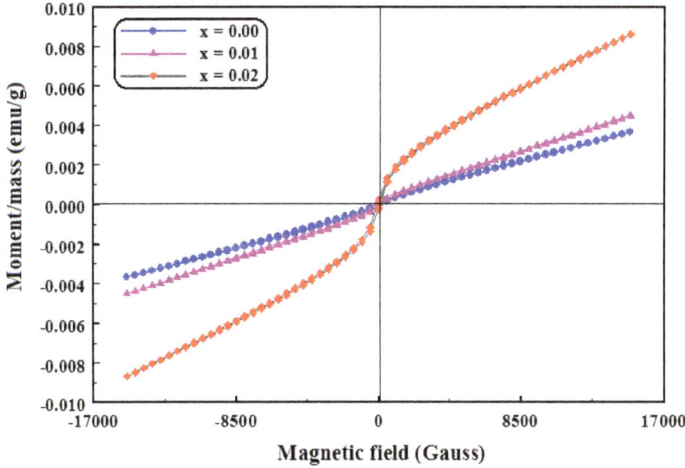

Figure 8: Magnetization curve for $Ni_{1-x}Fe_xO$.

Table 5. Parameters of $Ni_{1-x}Fe_xO$ from VSM Measurements.

Sample	Coercivity (G)	Magnetization (10^{-3} emu)	Retentivity (10^{-6} emu)
NiO	160.43	3.6816	73.363
$Ni_{0.99}Fe_{0.01}O$	108.31	4.4994	73.012
$Ni_{0.98}Fe_{0.02}O$	85.986	8.6438	234.00

4. CONCLUSION

In this Fe doped NiO samples, the Ni ions are partially substituted by the Fe ions, which are randomly localized over the host lattice. The cell parameters in table 1 show a decrease in the cell constant due to the addition of Fe. The MEM analysis shows that the doping of Fe in NiO changes the charge distribution in $Ni_{0.99}Fe_{0.01}O$, but when the doping concentration is increased no great changes in electronic properties are observed. Covalent bonding between Ni and Oxygen atoms is confirmed qualitatively and quantitatively. From the MEM, UV and VSM analysis the effect of doping could be observed only when Fe is less than 2%.

5. ACKNOWLEDGMENTS

Authors acknowledge with gratefulness SAIF, Cochin, Kerala, India for the XRD and UV data of the samples. The authors extend their thankfulness to The Madura College, Madurai, Tamil Nadu, India for the moral support.

REFERENCES

[1]   S. A. Wolf, D. DAwschalom, R.A. Buhrman, J.M. Daughton, S. Von Molnar, M.L. Roukes, A.Y. Chtchelkanova and D. M. Treger, Spintronics: a spin-based electronics vision for the future, Science 294 (2001) 1488-95.
        http://dx.doi.org/10.1126/science.1065389

[2]   T. Dietl, H. Ohno, F. Matsukura, J. Cibert and D. Ferrand, Zener model description of ferromagnetism in zinc-blende magnetic semiconductors, Science 287 (2000) 1019-22.
        http://dx.doi.org/10.1126/science.287.5455.1019

[3]   Yi-Dong Luo, Yuan-HuaLin, Xuehui Zhang, Deping Liu, Yang Shen, and Ce-Wen Nan, Ferromagnetic Behaviors in Fe-Doped NiO Nanofibers Synthesized by Electrospinning Method, Journal of Nanomaterials, 2 (2013) 525-93.

[4]   H.M. Rietveld, A profile refinement method for nuclear and magnetic structures, J. Appl. Crystallogr. 2 (1969) 65-71.
        http://dx.doi.org/10.1107/S0021889869006558

[5]   V. Petr˘ıc˘ek, M. Dus˘ek, L. Palatinus, in: JANA2000, The Crystallographic Computing System, Institute of Physics, Academy of Sciences of the Czech Republic, Praha, 2000.

[6]   A. D. Ruben, I. Fujio: Super-fast Program PRIMA for the Maximum-Entropy Method, Advanced materials Laboratory, National Institute for Materials Science, Ibaraki, Japan p. 305 (2004) 0044.

[7]   F.Izumi, R.A. Dilanian: Recent Research Developments in Physics, Part II, 3, Transworld, Research Network, Trivandrum, 2002, pp. 699–726.

[8]   K. Momma, F. Izumi, VESTA: a three-dimensional visualization system for electronic and structural analysis, J. Appl. Crystallogr. 41 (2008) 653.
        http://dx.doi.org/10.1107/S0021889808012016

CHAPTER 7

Synthesis and Characterization of Al_2TiO_5-TiO_2-Al_2O_3 Ceramics: Correlation with Charge Density

S.V. Meenakshi[1], R. Saravanan[2]

[1]Sri Meenakshi Government College for Women, Madurai- 625 002, Tamilnadu, India

[2]Research centre and PG Department of Physics, The Madura College, Madurai-625 011 Tamilnadu, India

Email: svmeenu74@yahoo.co.in; saragow@gmail.in

Abstract

Al_2TiO_5-TiO_2-Al_2O_3 ceramic material has been synthesized using solid state reaction at 1500 °C. The prepared ceramic sample was characterized using XRD and SEM. The charge density distribution of the ceramic has been constructed using the maximum entropy method (MEM). Using the electron density distribution, the nature of the bonding is discussed and is used to explain the properties of the synthesized ceramics. SEM images are used to find the aggregate particle size. The UV-Vis absorption spectrum and the dielectric characteristics are also done to discuss the nature of the sample.

Keywords

Al_2TiO_5, TiO_2, Al_2O_3, Ceramics, Charge Density

Contents

1.   INTRODUCTION

Aluminium titanate is an interesting ceramic material which has a high melting point (1860 °C) and a low thermal expansion coefficient ($1x10^{-6}$/ °C). The properties of low thermal expansion and high melting point make aluminium titanate a good shock resistant material. Aluminum titanate ceramic materials have many technological applications, among others, as thermal insulation liner, soot particulate filter in diesel engines, spacing rings of catalytic converters, in the foundry crucibles, launders, nozzles, riser tubes, pouring spouts and thermocouples for non-ferrous metallurgy and master moulds glass industries. Al_2TiO_5 is stable above 1280 °C and undergoes small decomposition within a temperature range of 900 – 1280 °C [1]. Al_2TiO_5-TiO_2-Al_2O_3 is a ceramic material consisting of a mixture of alumina (Al_2O_3) and titania (TiO_2) forming a solid solution with stoichiometric proportion of the components, Al_2O_3 and TiO_2. It is prepared by heating a mixture of alumina and titania to the temperature of 1500 °C, in air atmosphere. Aluminium titanate ceramics are doped usually with MgO, SiO_2 and ZrO_2 in order to stabilize the solid solution structure. Unfortunately, the crystal structure anisotropy that promotes the low thermal expansion provokes micro-cracking, prevents its technical use, as discussed in [1].Research activity on ceramics has enhanced the understanding of various properties like thermal expansion, thermal durability of the material, dielectric characteristics etc. Some disadvantages of the ceramic materials are brittleness and low workability (no ductility and low cutting performances). The ceramic brittleness is improved by the addition of fine particles of ZrO_2 to the Al_2O_3 matrix which increases fracture toughness by introducing many micro cracks at the boundaries [2]. Research on ductile ceramics [3] confirms thermal expansion by the addition of ZrO_2 [4]. Since, electron density distribution between the atoms reveal the nature of bonding, the maximum entropy method [5], MEM will be a perfect tool to explain the characteristics of the ceramic materials, and it has been used for the present analysis. The maximum entropy method is a powerful tool for studying electron density distributions including in low density regions. The software program PRIMA [6] was used for the MEM computations and the software program VESTA [7] was used for the 3D, 2D and 1D electron density distribution. MEM can very well be used for characterizing bonding electrons between atoms. The objective of the present work is to study the properties of the synthesized ceramic samples and to correlate them with the charge density using the MEM [5] technique.

2.   SAMPLE PREPARATION

Ceramic sample of Al_2TiO_5-TiO_2-Al_2O_3 was synthesized from α-Alumina (99.99%) and TiO_2 (99.99%) purchased from Alfa Aesar. Both the powders were mixed in the stoichiometric ratio and ground manually for one hour using an agate mortar. The ground powders were pressed into pellets and sintered in a tubular furnace for 3 hours at 1500°C. The process of grinding and pressing into pellets was repeated twice with sintering at 1350 °C for the duration of 8 hours and 3 hours. Finally, the pellets were ground for 2 hours and kept in the furnace for 4 hours at 1500 °C. The rate of increase in temperature of the furnace was 5 °C/min and the cooling rate was adjusted at 10 °C/min. The pellets were again made into fine powders to be able to record the XRD spectra.

3.   CHARACTERIZATION TECHNIQUES

3.1   X- RAY DIFFRACTION AND CHARGE DENSITY ANALYSIS

XRD data sets of Al_2TiO_5-TiO_2-Al_2O_3 sample were recorded at NIIST, Trivandrum, Kerala, India for phase identification and analysis. The X-ray scan was taken over an angular range of 10° to 120° with a step size of 0.02° and a counting time of 1.0 s at each step. Quantitative phase analysis was carried out using the Rietveld [8] pattern of fitting method JANA 2006 [9] and MAUD [10].The data was analyzed using JANA 2006 [9]. XRD profile and structural parameters like background, scale factors for all phases, asymmetry parameters, preferred orientation, unit cell parameters, atomic positional coordinates etc, were refined step by step avoiding correlations between parameters. There were three phases in the Al_2TiO_5-Al_2O_3-TiO_2 sample which were confirmed using JCPDS [11] to be responsible for the distorted orthogonal structure which was also confirmed by previous studies. The same XRD data was also refined using MAUD [10]. The structural parameters of the sample Al_2TiO_5-Al_2O_3-TiO_2 after refinement using JANA 2006 [9] are given in Table 1.

Table 1. Structural parameters of Al_2TiO_5- Al_2O_3-TiO_2.

Parameter	Al_2TiO_5	Al_2O_3	TiO_2
a (Å)	3.596821(65)	4.762373(913)	4.594762(167)
b (Å)	9.439894(196)	4.762373(0)	4.594762(0)
c (Å)	9.650063(167)	12.96824(532)	2.959011(532)
α (°)	90	90	90
β (°)	90	90	90
γ (°)	90	120	90
volume (Å)^3	327.6544(136)	254.7174(135)	62.47015(53)
ρ (g/cm^3)	3.685014(153)	3.986813(211)	4.245146(359)

The precise study of bonding in materials is always useful and inspiring. Since no two experimental data are identical, no study can ultimately manifest a legitimate picture. The bonding nature and the distribution of electrons in the bonding region can be anticipated using the MEM technique. The MEM refinements were carried out by dividing the unit cell into 64x64x64 pixels. For the numerical MEM computations, the software package PRIMA [6] was used. The MEM parameters of Al_2TiO_5 - TiO_2 - Al_2O_3 sample are given in table 2. In table 2, R_{MEM} represents the reliability index from the MEM refinement and wR_{MEM} represents the weighted reliability index. The software program VESTA [7] has been used for the analysis of the 3D, the 2D and the 1D electron density distribution. The XRD data used in the Rietveld refinement [8] was also analyzed by MAUD [10] and the XRD profile plot of Al_2TiO_5-TiO_2-Al_2O_3 powder after analysis using MAUD [10] is shown in figure 1.

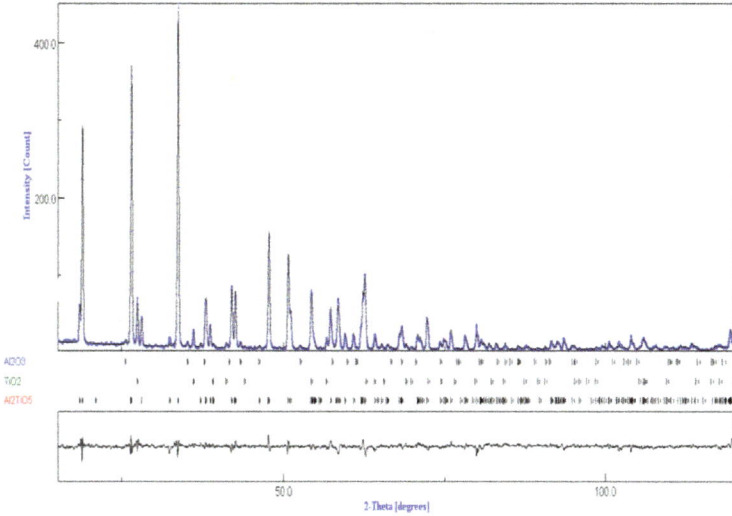

Figure 1. Refinement Profile using MAUD.

Table 2. Parameters of the MEM electron density analysis of Al_2TiO_5-Al_2O_3-TiO_2.

Parameter	Value
No. of cycles	1724
No. of electrons /unit cell	352
Lagrange parameter λ	0.005810
R_{MEM} (Reliability Index) (%)	2.9788
w R_{MEM} (Weighted Reliability Index) (%)	3.7375

The 3D electron density distribution given in fig 2. shows the distorted edge of shared oxygen (TiO_6) octahedral surrounding metal sites Ti1 and Ti2 with coordination number 6.

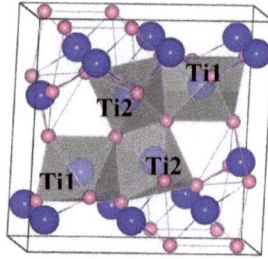

Figure 2. 3D Electron density distribution.

Figures 3a, 3b show both 3D and 2D electron density distribution for the bonding between the atoms Ti2 with O2. This type of bonding between a Ti (metal) atom and oxygen (non metal atom) being ionic is responsible for a high melting point and low thermal expansion of the Al_2TiO_5-TiO_2-Al_2O_3 ceramic sample.

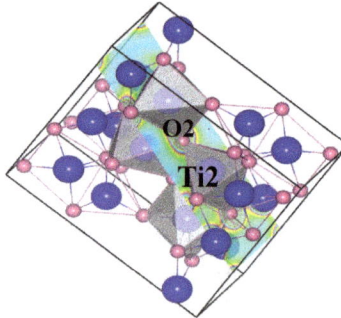

Figure 3a). 3D electron density distribution showing (012) plane.

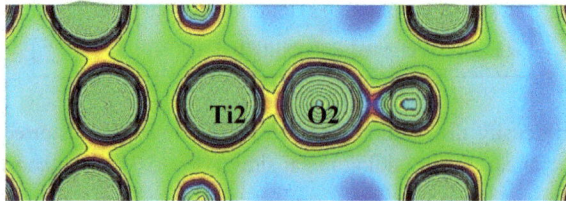

Figure 3b). 2D electron density distribution on (012) plane. Contour range is 0.5to 2 e/$\mathring{A}^3$ and contour interval is 0.1e/$\mathring{A}^3$.

This 1D electron density distribution profile shows the nature of bonding between a metal atom and a non metal atom. Figure 4 shows the interaction of a metal atom Ti2 with O1, O2 and O3. There exists ionic bonding between Ti with O1, O2 and O3.The electron densities at the mid distance between the atoms is given in table 3. The electron density at the mid distance between Ti2 and O2 seems high corresponding to less ionic, being responsible for a characteristic of a low melting point.

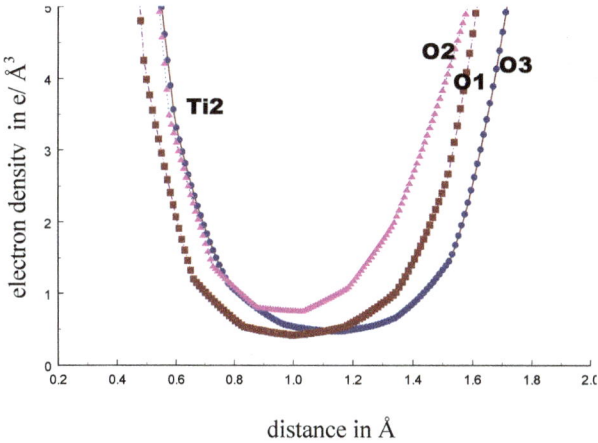

Figure 4. The 1D electron density profile showing bonding between Ti2 – O1, Ti2 – O2 and Ti2 – O3.

Table 3. The electron densities in the mid distance between the atoms.

Atom Pair	Distance in Å from first atom	Electron density (e/Å³)
Ti2-O1	1.0138	0.420
Ti2-O2	0.8420	0.913
Ti2-O3	1.0519	0.496

3.2   MICROSTRUCTURAL STUDY USING SEM

The microstructural analysis of the sintered samples was performed using scanning electron microscopy (SEM).The SEM images shown in figure 5 of the prepared sample reveals densification of the alumina matrix with some pores. Aggregate particle size (AGP) of Al_2TiO_5-TiO_2-Al_2O_3 is 19.670 µm in between the AGP of Al_2O_3 and TiO_2

confirms that the sample is well formed. No additional peaks are detected in the EDS spectrum of Al_2TiO_5-TiO_2-Al_2O_3, TiO_2 and Al_2O_3.

Al_2TiO_5

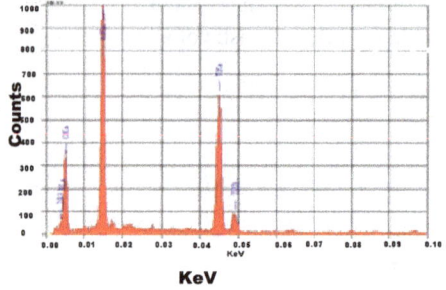

a) Al_2O_3

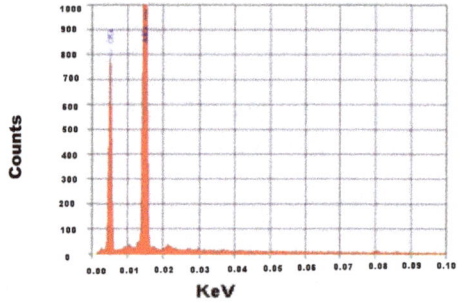

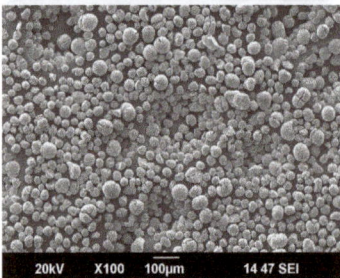
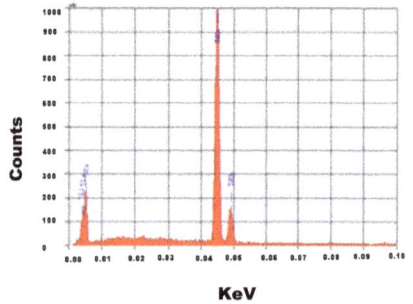

b) TiO_2

Figure 5. SEM and EDAX of Al_2TiO_5-TiO_2-Al_2O_3, TiO_2 and Al_2O_3.

3.3   OPTICAL MEASUREMENTS USING UV-VIS ABSORPTION SPECTRA

Optical absorption spectra were obtained at room temperature in the range of 100 – 2100 nm. Figure 6 shows the absorption band for the synthesized sample to be 100 nm in the visible region. For the lower wavelength, the absorption is high for the Al_2TiO_5-TiO_2-Al_2O_3 and also for Al_2O_3, TiO_2. This indicates a high band gap, which is the characteristic of an insulator. High frequency absorption indicates the nature of ionic bonding which is again related with a high band gap. Optical measurements show a high absorption band in the visible as well as in the UV region, which again confirms a high band gap.

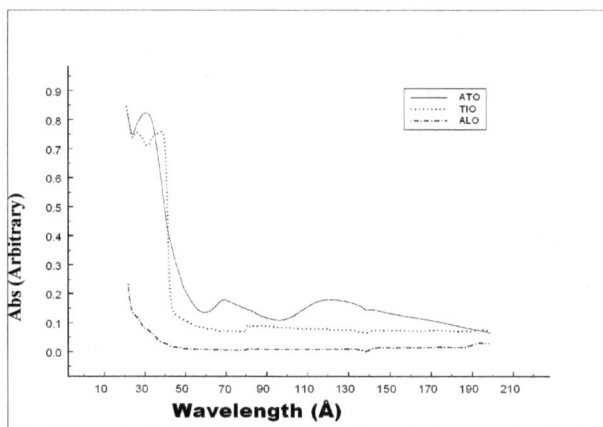

Figure 6. UV-Vis Absorption spectrum.

3.4   DIELECTRIC STUDIES

The dielectric study was done by observing the variations in the permittivity with the frequency at room temperature. Figure 7 shows the frequency vs. permittivity graph. For higher frequencies, the permittivity is low and it takes a larger value in the lower frequency range. This confirms the nature of capacitive behavior. A dielectric is usually used to increase the capacitance. As the frequency increases, the value of the capacitance decreases due to the decrease in the permittivity. In dielectric studies, the permittivity is high for the low frequencies and at the most zero at high frequencies. This justifies the relation between the dielectric constant and relative permittivity ($k = \epsilon_r$).

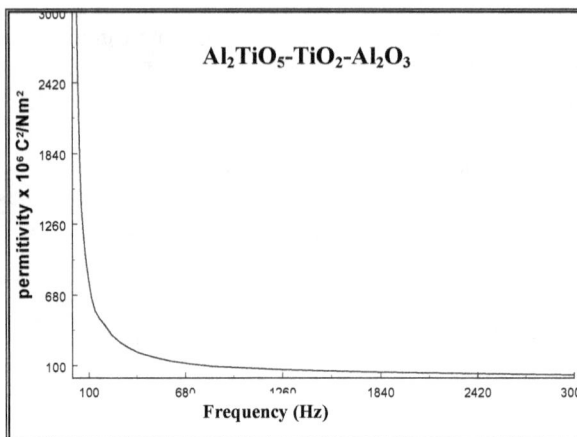

Figure 7. Frequency vs. permittivity graph.

4. CONCLUSION

The structure of a Al_2TiO_5-TiO_2-Al_2O_3 solid solution is analyzed using the XRD powder refinements. The electron density distribution in 1D, 2D and 3D is studied using the MEM analysis. The bonding between Ti, O is ionic. The strength of bond depends on the sizes of the atoms which form the bond and also the electrons involved in bonding. This causes the different characteristics of ceramic materials (high melting point, low thermal expansion). The distorted structure with less electron density for the covalent bond between the oxygen atoms may possibly be responsible for the brittleness of the ceramic. The high band gap energy confirms ionic bonding and the dielectric studies reveal the poor electrical conductivity with good capacitive property.

ACKNOWLEDGEMENTS

The authors acknowledge with gratefulness Dr Viswa Bhaskaran, the M.D, and Mr.Raja of V.B. Ceramics, Chennai, Tamilnadu, India for their help rendered, NIIST, Trivandrum, Kerala, India for the XRD spectra, SAIF- KOCHI, cochin, Kerala, India, for the SEM-EDS of all the samples. The authors extend their Thankfulness to The Madura College and Sri Meenakshi Government Arts College for Women, Madurai, Tamilnadu, India for the moral support. The authors also thank Mr. Sasikumar Research Scholar, The Madura College, for his timely help.

REFERENCES

[1]   R.D. Skala, D. Li. L.M. Low Diffraction, Structure and phase stability studies on Aluminium titanate J. European ceramic society 29 (2009) 67-75. *http://dx.doi.org/10.1016/j.jeurceramsoc.2008.05.037*

[2]   N. Claussen, J. Steeb, R.F. Pabst. Am. ceram. Soc. Bull. 56 (1977) 559-569.

[3]   T. Shimazu, M. Miura, N. Ishu, T. Ogawa, K. Ota, H. Maeda, E.H. Ishida. J. msea. 487 (2008) 340-346. *http://dx.doi.org/10.1016/j.msea.2007.10.026*

[4]   Hyung chul kim, Kaesung Lee, Ohseong kweon, Christos G.Aneziris, I.K.Jin kim. J. Eur. Ceram. Soc. 27 (2007) 1431-1434. *http://dx.doi.org/10.1016/j.jeurceramsoc.2006.04.024*

[5]   S. Kumazawa, Y Kubota, M. Takata, M.Sakata and Y.Ishibashi J. Appl. Cratallogr.26 (1993)453. *http://dx.doi.org/10.1107/S0021889892012883*

[6]   Izumi F and Dilanian R.A, Recent Research Developments in Physics Vol 3 Part II (Trivandrum Transworld Research Network) (2002) 699-726.

[7]   Momma K and Izumi F Commission on Crystallographic Computing, IUCr News letter 7 (2006) 106.

[8]   H.M. Reitveld, J. Appl. Crystallogr. 2 (1969) 65-71. *http://dx.doi.org/10.1107/S0021889869006558*

[9]   V. Petricek, M. Ducek and L. palatines JANA 2006. The crystallographic computing system, Institute of Physics, Praha, Czech Republic (2006).

[10]   Luca Lutterotti Instrumental broadening determination Universita' di Trento 38050 Trento, Italy (2006).

[11]   JCPDS CAS No. 26-0040.

[12]   T. Hahn, Netherlands (The international Union of Crystallography, Springer), Space Group Symmetry, International Tables for Crystallography, Vol. A, 5th ed., edited by T. Hahn (2005) 584-585.

[13]   R. Saravanan Practical application of Maximum entropy method in electron density and studies Phys. Scr. 79 (2009) 048303. *http://dx.doi.org/10.1088/0031-8949/79/04/048303*

[14]   D M Collins 1982 Electron density images from imperfect data by iterative entropy maximization Nature. 298 49. *http://dx.doi.org/10.1038/298049a0*

[15] W. Jauch The maximum entropy method in charge density studies. II. General aspects of reliability Acta Crystallogr. A 50 (1994) 650.
http://dx.doi.org/10.1107/S0108767394004472

[16] A.K. Livesey and J Skilling Maximum entropy estimate of electron density function Acta Crystallogr. A 41(1985) 232.
http://dx.doi.org/10.1107/S0108767385000526

[17] L. Giordano, M. Viviani, C. Bottino, M.T. Buscaglia, V. Buscaglia, P. Nanni. J Eur. Ceram. Soc. 22 (2002) 1811-1822.
http://dx.doi.org/10.1016/S0955-2219(01)00503-9

[18] P. Scardi, L. Lutterotti, Maistrelli, Powder diffr., 9(3) (1994) 180-186.
http://dx.doi.org/10.1017/S0885715600019187

[19] R.A. Young, editor, The Rietveld method, IUCr, Oxford university Press, Oxford, UK, 1993.

[20] Davor Balzar, Nicolac C.Popa Analyzing microstructure by Rietveld refinement. J. Rigaku 22(1) (2005) 16-25.

CHAPTER 8

Electronic Charge Density Distributions in Sb$_2$O$_3$

T. K. Thirumalaisamy[1], S. Saravanakumar[2], R. Saravanan[3]

[1]Department of Physics, H.K.R.H. College, Uthamapalayam-625 533, Tamil Nadu, India

[2]Department of Physics, Kalasalingam University, Krishnankoil, Viruthunagar - 626 126, India

[3]Research Centre and PG Department of Physics, The Madura College, Madurai-625 011, Tamil Nadu, India

Email: tktsamy67@gmail.com; saragow@gmail.com; saravanaphysics@gmail.com

Abstract

High-resolution charge density distribution maps have been elucidated using the maximum entropy method (MEM) for the cubic polymorph of antimony oxide (Sb$_2$O$_3$) using experimental X-ray structure factors. Information about the nature of bonding and relative bond strengths in the cubic polymorph of antimony oxide (Sb$_2$O$_3$) were extracted. The microstructure and the band gap energy of antimony oxide (Sb$_2$O$_3$) have been studied through scanning electron microscopy and UV-Visible analysis respectively.

Keywords

Sb$_2$O$_3$, Charge Density, MEM, XRD, SEM, UV-Vis.

Contents

1.   INTRODUCTION

Nowadays, antimony oxide based glasses are of great importance in the field of fundamental research and technical applications [1-3]. Mostly, three different kinds of antimony oxides Sb_2O_3 [4], Sb_2O_4 [5] and Sb_2O_5 [6] are known. Antimony oxide (Sb_2O_3) exists in two crystalline phases, valentinite (orthorhombic phase) [7] and senarmontite (cubic phase) [8]. Orman et al [9] have confirmed the existence of thermal phase transition between low temperature cubic phase and high temperature orthorhombic phase in antimony oxide (Sb_2O_3). Sahoo et al. [10] have reported some potential applications of these materials related to its high refractive index. Lu et al. [11] have proposed that modified nanorods involving antimony oxide (Sb_2O_3) may be suitable for biomedical sensing.

In the literature, a lot of reports are available on the fabrication of transparent crystallized glasses consisting of non-linear optical nanocrystals [12-14]. Satyanarayana et al., [15] have demonstrated that the antimony oxide participates in the glass network with SbO_3 structural units and can be viewed as tetrahedrons with the oxygen atoms situated at three corners and the lone pair of electrons of antimony (Sb^{3+}) at the fourth corner localized in the third equatorial direction of Sb atom. The deformation of this pair probably plays a role in the glass ceramics for exhibit non-linear optical susceptibility. Falcao Filho et al. [16] have reported the measurements of the infrared third-order non-linear optical response of new glasses with different compositions of antimony oxide (Sb_2O_3). Furthermore, Deng et al. [17] have established that antimony oxide (Sb_2O_3) is an interesting material exhibiting unique optical, electronic and optoelectronic properties.

Most of the works on the synthesized nanostructures belong to the orthorhombic phase [17-21], while few works refer to the cubic antimony oxide (Sb_2O_3) nanostructures [22, 23]. Various methods, such as chemical methods [24, 25], vapour transport and condensation [24], and carbothermal reduction [21], have been used for the growth of antimony oxide (Sb_2O_3) one-dimensional nanostructures, nanoparticles and hierarchical nanostructures respectively.

Understanding chemical and physical properties of any material requires the knowledge of their electronic charge density distribution. The objective of the present study can be considered to be an attempt in visualizing the electronic charge density distribution and bonding nature in antimony oxide (Sb_2O_3). In the present work, precisely collected powder X-ray intensity data of antimony oxide (Sb_2O_3) has been refined using the Rietveld [26] method. The maximum entropy method (MEM) [27] has been introduced in charge density reconstruction at any position within the unit cell. The size of the particle and the band gap energy of antimony oxide (Sb_2O_3) have also been estimated through scanning electron microscope (SEM) and UV–visible analysis respectively.

2. EXPERIMENTAL ANALYSIS

2.1 XRD POWDER DATA COLLECTION AND STRUCTURAL REFINEMENT

The high purity analytical grade (Alfa Aesar, 99.995%) antimony oxide (Sb_2O_3) powder purchased from scientific suppliers has been used for the present analysis. The X-ray powder diffraction (XRPD) technique was used to investigate the inner arrangement of atoms or molecules in a crystalline material. The X-ray powder diffraction (XRPD) measurements in this work were done at the International Research Center, Kalasalingam University, Krishnan Koil, Virudunagar, India, using an Bruker - D8 Advance ECO XRCD systems (Germany), X-ray diffractometer using pure Cu–Kα_l radiation. The wavelength used for the X-ray intensity data collection was 1.5406 Å. The accelerating voltage and the applied current density were 40 KV and 20 mA/cm^2 respectively. Measurements were scanned with a 2θ range from 10° to 120° with a 0.0203° step size, using a step time of 65.6 s. The observed X-ray profile was compared with JCPDS (Joint Committee Powder Diffraction Standards). All X-ray peaks are well matched with the observed X-ray profile and it has a cubic structure of antimony oxide (Sb_2O_3) with the space group $Fd\bar{3}m$ (227). All peaks are indexed and shown in Fig.1. The structure for Sb_2O_3 is identified as the one with cell parameters a=b=c=11.15 Å and the initial atomic coordinates for Sb and O are given as (0.8852, 0.8852, 0.8852) and (0.1865, 0, 0) respectively.

Scanning electron microscope (SEM) micrographs of antimony oxide (Sb_2O_3) were obtained under different magnifications on a field emission SEM apparatus (JSM-6390LV, JEOL) operating at acceleration voltage of 30 kV, one of which is shown in Fig. 2. The surface morphology of Sb_2O_3 is clearly visible in Fig. 2 and the particle size of the observed system is found to be around 0.6 μm. Energy dispersive spectroscopy (EDS) measurements of antimony oxide (Sb_2O_3) were performed using a scanning electron field microscope (JED–2300, JEOL). The EDS spectrum in Fig. 3 shows the presence of appropriate quantities of Sb and O elements. No additional impurities are detected in the EDS spectrum of antimony oxide (Sb_2O_3). The peaks indicate the presence of antimony (Sb) and oxygen (O) atoms. The sharp intense peaks on the X-ray powder diffraction (XRPD) pattern reveal that the crystallites are pure and dislocation free.

The well-known Rietveld [26] method is a method for refining crystal structure from powder diffraction profile and it gives much structural information. In this work, the

Rietveld [26] analysis was performed for the data set using the software package JANA 2006 [28]. The fitted profile of the observed and calculated relative X-ray intensities along with their differences for antimony oxide (Sb_2O_3) is shown in Fig. 4. The refined fractional atomic coordinates and thermal vibration parameters with the corresponding standard deviations in parenthesis are given in table 1. The thermal parameter for the Sb atom is much less than that of the O atom due to less atomic weight of O atom. The observed and calculated structure factors from the powder data refinement of antimony oxide (Sb_2O_3) using JANA 2006 [28] are presented in table 2 and were used to construct the electron density distribution in the unit cell. The refined structural parameters of the chosen material from the Rietveld [26] refinement of powder XRD are presented in table 3. In this table R_p, $_wR_p$, R_{obs} and GOF represent the reliability indices for profile, weighted reliability indices for profile, reliability indices for the observed structure factors and goodness of fit respectively.

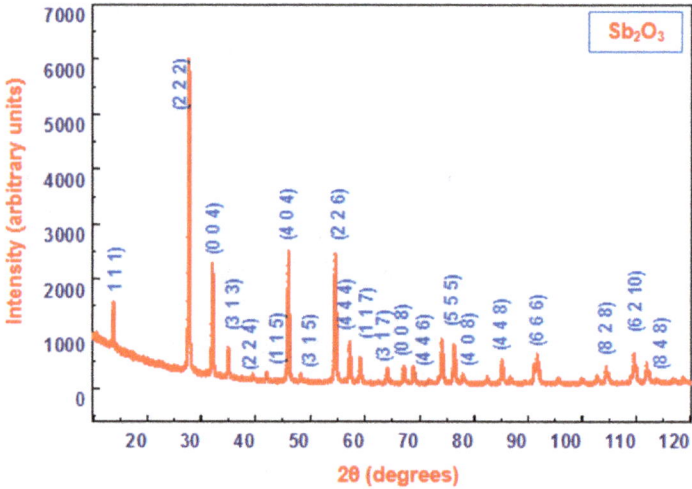

Figure 1. Raw X-ray powder diffraction profile of antimony oxide (Sb_2O_3).

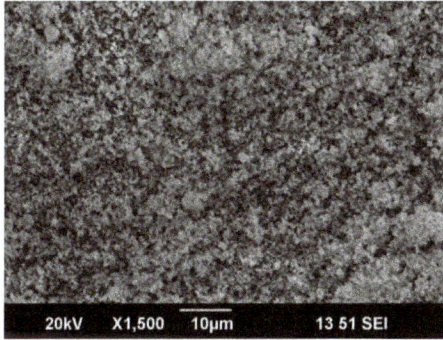

Figure 2. SEM micrograph of antimony oxide (Sb_2O_3).

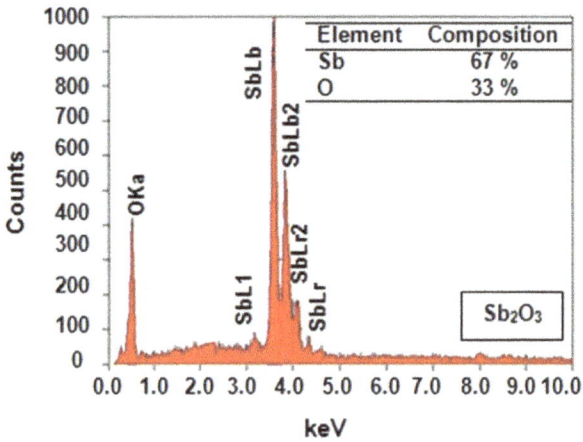

Figure 3. EDS spectrum of antimony oxide (Sb_2O_3).

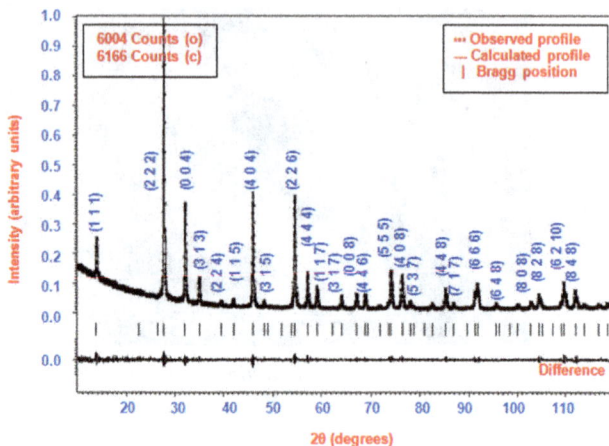

Figure 4. Fitted powder XRD profile of antimony oxide (Sb_2O_3). (The vertical line shows the Bragg peaks. The difference curve shows the difference between the observed and calculated profiles).

2.2 UV-VISIBLE ANALYSIS

The UV-visible analysis is a process in which the outer electrons of atoms or molecules absorb radiant energy and undergo transitions to high energy levels. In this process, the spectrum obtained due to optical absorption can be analyzed to get the optical band gap energy of the material. Fig. 5 shows the UV–Vis absorption spectrum of antimony oxide (Sb_2O_3) measured at room temperature in the wavelength ranging from 200 to 2500 nm. The optical band gap energy (E_g) was evaluated from the absorption spectra and the optical absorption coefficient (α) near the absorption edge according to the equation proposed by Wood and Tauc [29]. The band gap of the antimony oxide (Sb_2O_3) was estimated by plotting $(\alpha h v)^2$ versus hv and extrapolating the linear portion near the onset of absorption edge to the energy axis. The reported optical band gap energy values of antimony oxide (Sb_2O_3) are 3.3 eV [29] and 4.0 eV [30] for orthorhombic and cubic crystal forms, respectively. Tigau et al. [31] measured optical response in the energy region up to 4.5 eV, while Validzic et al [32] reported the optical absorption edge energy from 4.1 to 4.4 eV in their reflectance measurements respectively. In the present work, the value of the optical band gap energy obtained is 4.19 eV and agrees well with the earlier reported values.

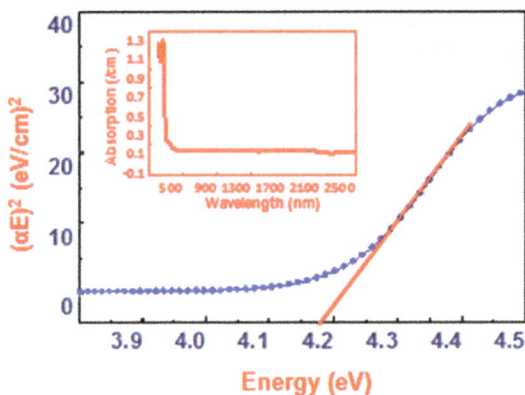

Figure 5. UV-Vis spectrum of antimony oxide (Sb$_2$O$_3$).

2.3 CHARGE DENSITY DISTRIBUTION STUDY USING MAXIMUM ENTROPY METHOD (MEM)

The maximum entropy method (MEM) [27] is a versatile tool to determine the charge density distribution in a unit cell. It gives an accurate charge density and is more reliable than other methods such as Fourier method for finding charge density distribution in crystallography. The resultant charge density from the maximum entropy method (MEM) is plotted with the help of the visualization software VESTA [33]. In the present work, the MEM [27] refinement was carried out by dividing the unit cell into 128×128×128 pixels. The uniform prior density was determined by dividing the total number of electrons by the volume of the unit cell. The obtained parameters from charge density refinement of antimony oxide (Sb$_2$O$_3$) have been given in table 4. In this table, R$_{MEM}$ represents the reliability index from the MEM [27] refinements and wR$_{MEM}$ represents the weighted reliability index.

Fig. 6a shows the representation of atoms in the unit cell antimony oxide (Sb$_2$O$_3$). Fig. 6b shows the three-dimensional charge density distribution of antimony oxide (Sb$_2$O$_3$) represented in the unit cell considering the iso-surface level of 1.1 e/Å3. The O atoms in the Sb$_4$O$_6$ structural units form vertices of an octahedron with a 2.936 Å edge. Each Sb atom bonded with three O atoms at 1.968 Å. Each oxygen atom is shared by two Sb atoms. A tetrahedral arrangement around Sb is completed by its 5s^2 lone-pair electrons. The Sb$_4$O$_6$ structural unit has tetrahedral symmetry and could be described as a large tetrahedron with four lone-pairs as corners. These tetrahedra are arranged face to face in such a way that the O atoms of two adjacent faces are corners of the slightly distorted octahedron. The $O - O$ distances in the Sb_4O_6 structural unit is 2.922 Å. Different views of the same in one-fourth of the unit cell and one-eighth of unit cell of antimony oxide (Sb_2O_3) with the iso-surface level of 1.0 $e/Å^3$ are shown in Fig. 7a and b respectively. The accumulation of charges is around the atoms and it is towards the charge center. This figure shows that the spatial charges are distributed along the bonding directions between Sb and O atoms. Fig. 7c represents the unit cell of antimony oxide (Sb_2O_3) consisting of all the distorted octahedral arrangements and the view in one-eighth of the unit cell is shown in Fig. 7d. A 3D view of all the Sb_4O_6 structural units in the unit cell and one-eighth of the unit cell were presented in Fig. 7e and f respectively. Another view of a single Sb_4O_6 structural unit in the unit cell of antimony oxide (Sb_2O_3) without the iso-surface level is shown in the Fig. 7g and the same with the iso-surface level of 1.0 e/Å3 is shown in the Fig. 7h. Fig. 8a shows the (101) plane passing through Sb and O (at x + $\frac{1}{4}$, $-z + \frac{1}{4}$, y + $\frac{1}{4}$). The 2D charge density map on the (101) plane passing through Sb and O (at x + $\frac{1}{4}$, $-z + \frac{1}{4}$, y + $\frac{1}{4}$) is shown in the Fig. 8b (contour range is from 0 $e/Å^3$ to 4 $e/Å^3$ and contour interval is 0.15 $e/Å^3$). The spherical nature of the atoms of Sb and O and the accumulation of charges towards the charge center was clearly visible in this (101) plane. The electronegativity for Sb and O atoms are 2.05 and 3.44 respectively. The difference of electronegativity leads to polar covalent bonding between the Sb and the O atom.

a)   b)

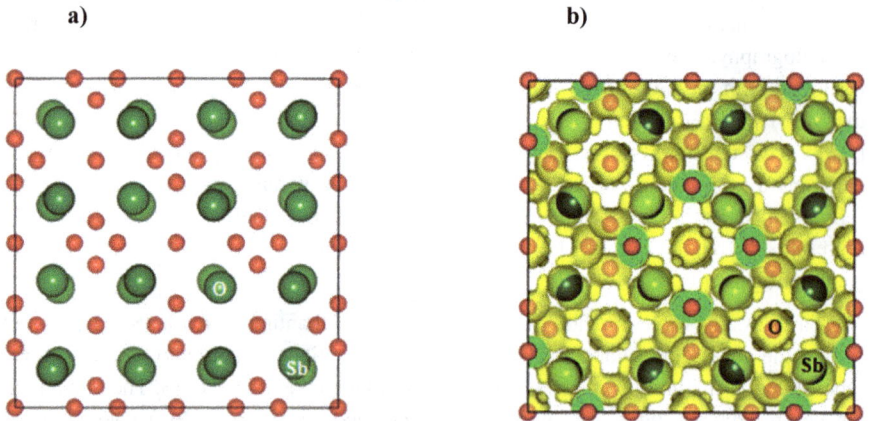

Figure 6. (a) The representation of atoms in the unit cell of antimony oxide (Sb₂O₃); (b) Corresponding 3D charge density distribution considering the iso-surface level of 1.0 e/Å³.

a)

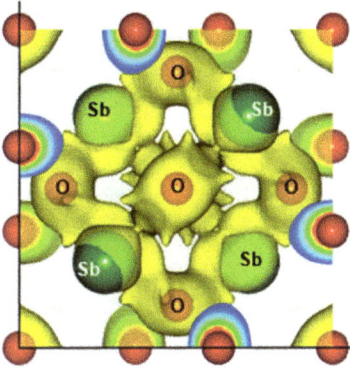

b)

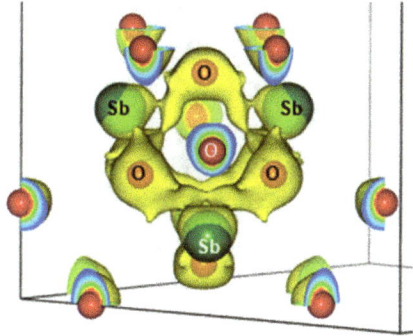

c)

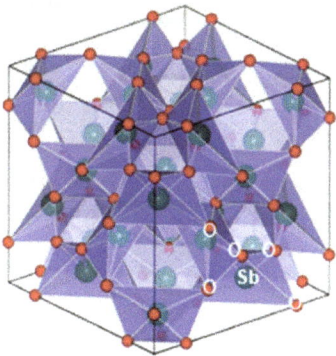

d)

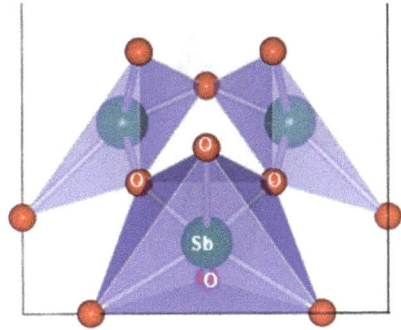

e)

f)

g)

h)

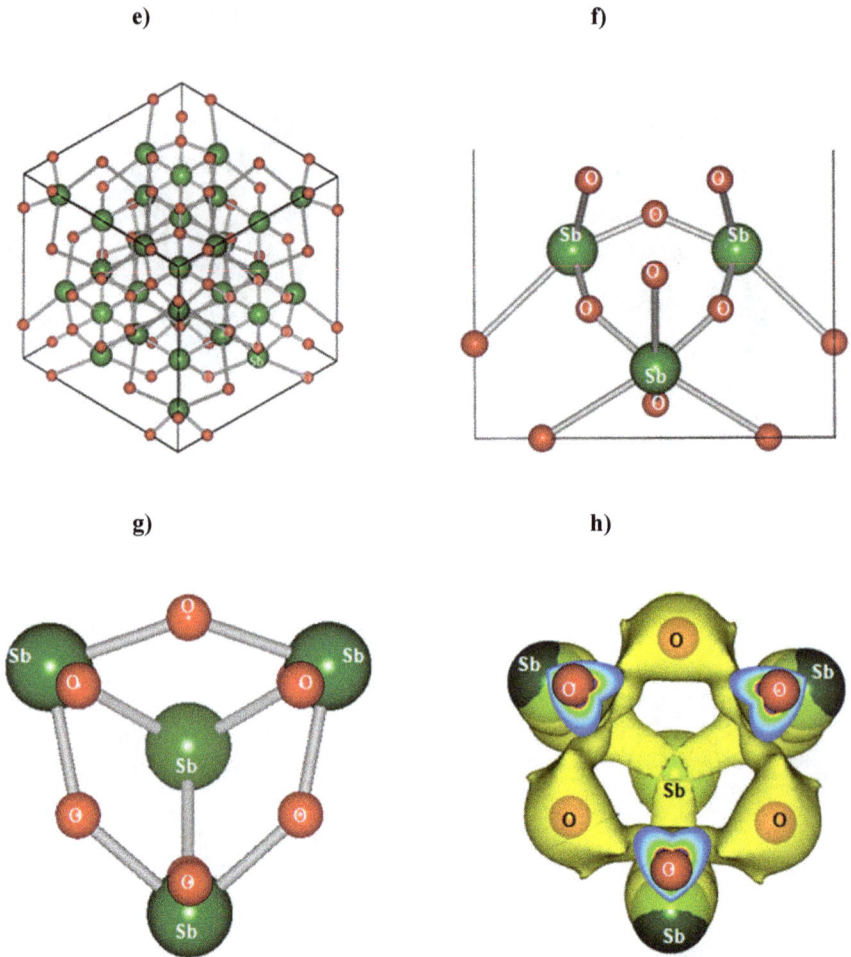

Figure 7. (a) View of a One-fourth of the unit cell (iso-surface level of 1.0 e/$\mathring{A}^3$); (b) One-eighth of the unit cell (iso-surface level of 1.0 e/$\mathring{A}^3$); (c) All the distorted octahedral arrangements in the unit cell; (d) Distorted octahedral arrangement in one-eighth of the unit cell; (e) All the Sb_4O_6 structural unit in the unit cell; (f) Sb_4O_6 structural unit in one-eighth of the unit cell; (g) Single Sb_4O_6 structural unit without iso-surface level; (h) Single Sb_4O_6 structural unit without iso-surface level of 1.0 e/$\mathring{A}^3$.

The one dimensional charge density profile variation plotted between

$Sb - O$ (at $x + \frac{1}{4}$, $-z + \frac{1}{4}$, $y + \frac{1}{4}$)

and $O - O$ are shown in Fig. 9a and b respectively. The mid bond positions and the electron densities between the atoms are presented in table 5. In all the 1D electron density profiles of Sb2O3 the first atom has been considered to be in the origin. The value of mid bond electron density between $Sb - O$ is found to be 0.9272 e/ Å3 at a distance of 0.8996 Å. This value typically shows the bonding to be polar covalent.

a)   b)

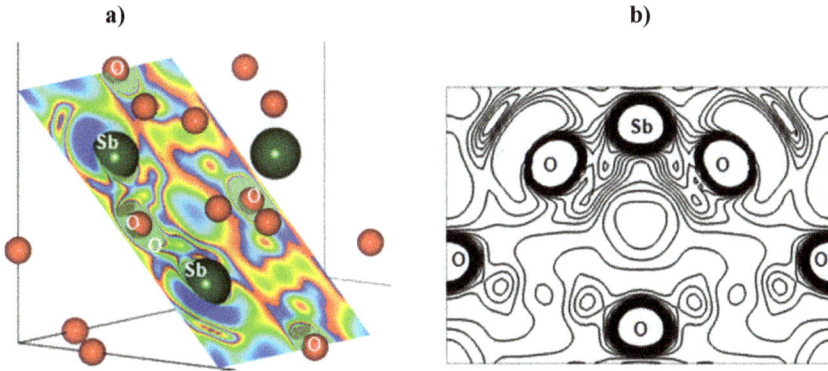

Figure 8. (a) The (101) plane passing through Sb and O (at $x + \frac{1}{4}, -z + \frac{1}{4}, y + \frac{1}{4}$) of antimony oxide (Sb$_2O_3$) (one-fourth of the unit cell); (b) The corresponding 2D MEM charge density.

a)   b)

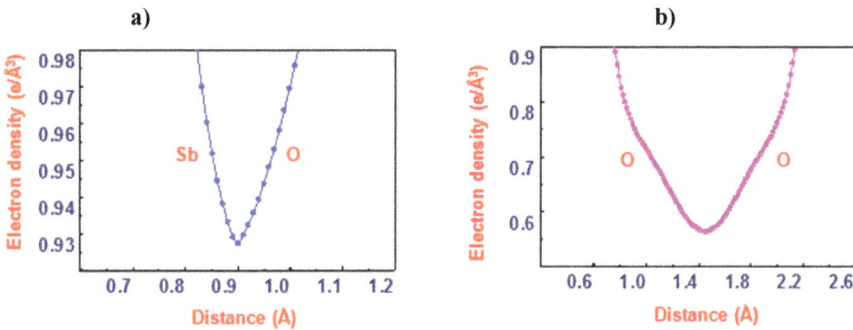

Figure 9. (a) 1D variation of electron density in the unit cell of Sb$_2$O$_3$ along Sb – O bond; (b) O – O bond.

3.   CONCLUSION

The cubic polymorph of antimony oxide (Sb_2O_3) has been characterized by X-ray diffractometer for structural analysis, scanning electron microscopy for morphological analysis, Energy dispersive analysis using X-ray for the elemental compositional analysis and the UV-Visible spectrophotometer for the optical analysis. The structure factors for Sb_2O_3 have been extracted using the Rietveld refinement technique. The resultant structure factors are used to construct the charge density in the unit cell using the MEM technique. The bonding behavior of Sb_2O_3 has been analyzed using the charge density distribution and is found to be polar covalent.

REFERENCES

[1]   G. Poirier, M. Poulain and M. Poulain, Copper and lead halogeno-antimoniate glasses, J. Non-Cryst. Solids, 284 (2001) 117-122.
http://dx.doi.org/10.1016/S0022-3093(01)00389-1

[2]   M. Nalin, M. Poulain, M. Poulain, S. J. L. Ribeiro and Y. Messaddeq, Antimony oxide based glasses, J. Non-Cryst. Solids, 284 (2001) 110-116.
http://dx.doi.org/10.1016/S0022-3093(01)00388-X

[3]   R.E. Araujo, C.B. Araujo, G. Poirier, M. Poulain and Y. Messaddeq, Nonlinear optical absorption of antimony and lead oxyhalide glasses, Appl. Phys. Lett., 81 (2002) 4694-4696.
http://dx.doi.org/10.1063/1.1529310

[4]   Z.T. Deng, D. Chen, F.Q. Tang, X.W. Meng, J. Ren and L. Zhang, *Orientated attachment* assisted self-assembly of Sb_2O_3 nanorods and nanowires: End-to-end versus side-by-side, Journal of Physical Chemisrty C, 111 (2007) 5325.
http://dx.doi.org/10.1021/jp068545o

[5]   Z.J. Zhang and X.Y. Chen, Biomolecule-assisted hydrothermal synthesis of Sb_2S_3 and Bi_2S_3 nanocrystals and their elevated-temperature oxidation behavior for conversion into α-Sb_2O_4 and Bi_2O_3, J. Phy. Chem. Solid, 70 (2009) 1121.
http://dx.doi.org/10.1016/j.jpcs.2009.06.010

[6]   L. Guo, Z. Wu, T. Liu, W. Wang and H. Zhu, Synthesis of novel Sb_2O_3 and Sb_2O_5 nanorods, Chem. Phy. Lett. 318 (2000) 49-52.
http://dx.doi.org/10.1016/S0009-2614(99)01461-X

[7]   C. Svensonn, The crystal structure of orthorhombic antimony trioxide, Sb_2O_3, Acta Cryst. B, 301974 (1974) 458461.
http://dx.doi.org/10.1107/s0567740874002986

[8] C. Svensonn, Refinement of the crystal structure of cubic antimony trioxide, Sb_2O_3, Acta Cryst. B, 31 (1975) 2016-2018.
http://dx.doi.org/10.1107/S0567740875006759

[9] R.G. Orman and D. Holland, Thermal phase transitions in antimony (III) oxides, J. Solid State Chem., 180 (2007) 2587-2596.
http://dx.doi.org/10.1016/j.jssc.2007.07.004

[10] N.K. Sahoo and K.V.S.R. Apparao, *Process-parameter optimization* of Sb_2O_3 films in the ultraviolet and *visible region for interferometric applications,* Appl. Phys. A, 63 (1996) 195-202.
http://dx.doi.org/10.1007/bf01567650

[11] X. Lu, Z. Wen and J. Li, Hydroxyl-containing antimony oxide bromide nanorods combined with chitosan for biosensors, Biomaterials, 27 (2006) 5740-7
http://dx.doi.org/10.1016/j.biomaterials.2006.07.026

[12] R. Sakai, Y. Benino and T. Komatsu, *Enhanced second harmonic generation* at surface in transparent nanocrystalline TeO_2-based glass ceramics, Appl. Phys. Lett., 77 (2000) 2118-2120.
http://dx.doi.org/10.1063/1.1313805

[13] G. Senthil, K.B.R. Varma, Y. Takahashi and T. Komatsu, Nonlinear-optic and ferroelectric behavior of lithium borate–strontium bismuth tantalate glass–ceramic composite, Appl. Phys. Lett., 78 (2001) 4019-4021.
http://dx.doi.org/10.1063/1.1380237

[14] Y. Takahashi, K. Kitamura, Y. Benino, T. Fujiwara and T. Komatsu, luminescent properties of $CaAl_2Si_2O_8$ nano crystallized glass, Appl. Phys. Lett., 86 (2005) 091110.
http://dx.doi.org/10.1063/1.1879114

[15] T. Satyanarayana, I.V. Kityk, M. Piasecki, P. Bragiel, M.G. Brik, Y. Gandhi and N.Veeraiah, Structural investigations on $PbO–Sb2O3–B2O3\_CoO$ glass ceramics by means of spectroscopic and dielectric studies, J. Phys.: Condens. Matter., 21 (2006) 245104.
http://dx.doi.org/10.1088/0953-8984/21/24/245104

[16] E.L. Falcao-Filho, C.B. de Araújo, G.S. Bosco, Maciel, L.H. Acioli, M. Nalin, and Y. Messaddeq, Antimony orthophosphate glasses with large nonlinear refractive-indices, low two photon absorption coefficients, and ultrafast response, J. Appl. Phys., 97 (2005) 013505.
http://dx.doi.org/10.1063/1.1828216

[17]   Z.T. Deng, F.Q. Tang, D. Chen, X.W. Meng, L. Cao and B.S. Zou, *A simple solution route to single-crystalline Sb₂O₃ nanowires* with rectangular cross sections, J. Phys. Chem. B, 110 (2006) 18225.
        http://dx.doi.org/10.1021/jp063748y

[18]   Y. Zhang, G. Li, J. Zhang and L. Zhang, Shape-controlled growth of one-dimensional Sb₂O₃ nanomaterials, Nanotechnology, 15 (2004) 762-765.
        http://dx.doi.org/10.1088/0957-4484/15/7/007

[19]   Z. Deng, D. Chen, F. Tang, J. Ren and A.J. Muscat, S*ynthesis* and *purple-blue* emission of antimony trioxide single-crystalline *nanobelts* with *elliptical cross section,* Nano Res., 2 (2009) 151-160.
        http://dx.doi.org/10.1007/s12274-009-9014-y

[20]   Q. Wang, S. Ge, Q. Shao and Y. Zhao, *Self-assembly* of *Sb₂O₃ nanowires into microspheres*: *synthesis* and *characterization,* Phys. B, 406 (2011) 731-736.
        http://dx.doi.org/10.1016/j.physb.2010.11.038

[21]   G.Fan, Z. Huang, C. Chai and D. Liao, *Synthesis* of *micro-sized* Sb₂O₃ *hierarchical structures* by *carbothermal reduction method,* Mater. Lett., 65 (2011) 1141-1144.
        http://dx.doi.org/10.1016/j.matlet.2010.09.084

[22]   T. Som and B. Karmakar, *Structure* and *properties* of *low-phonon antimony glasses* and *nano glass-ceramics* in *K₂O*-B₂O₃-Sb₂O₃ system, J Non-Cryst. Solids, 356 (2010) 987-999.
        http://dx.doi.org/10.1016/j.jnoncrysol.2010.01.026

[23]   Xu C.H, Shi S.Q, Surya C and Woo C.H, Synthesis of antimony oxide nanoparticles by vapor transport and condensation, J. Mater. Sci., 42 (2007) 9855-9858.
        http://dx.doi.org/10.1007/s10853-007-1799-z

[24]   Ye C, Wang G, Kong M and Zhang L, Controlled Synthesis of Sb₂O₃ Nanoparticles, Nanowires, and Nanoribbons, J. Nanomater. 2006, 95670 p.1-5.
        http://dx.doi.org/10.1155/JNM/2006/95670

[25]   Li Y, Zhang Y.X, Fang X.S, Zhai T.Y, Liao M.Y, Wang H.Q, Li G.H, Koide Y, Bando Y and Goldberg D, *Sb₂O₃* nanobelt networks for excellent visible-light-range photodetectors, Nanotechnolog, 22 (2011) 165704.
        http://dx.doi.org/10.1088/0957-4484/22/16/165704

[26]   Rietveld H.M, A Profile refinement method for nuclear and magnetic structures. J. Appl. Crystallogr., 2 (1969) 65-71.
        http://dx.doi.org/10.1107/S0021889869006558

[27]   Collins D.M, Electron density images from imperfect data by iterative entropy maximization, Nature, 298 (1982) 49-51.
http://dx.doi.org/10.1038/298049a0

[28]   Petricek V, Dusek M and Palatinus L, JANA 2006, The Crystallographic Computing System, Institute of Physics, Academy of Sciences of the Czech Republic, Praha, 2000.

[29]   Wood D. L and Tauc J, Weak absorption tails in amorphous semiconductors Phys. Rev., B5 (1972) 3144.
http://dx.doi.org/10.1103/PhysRevB.5.3144

[30]   Wolffing B and Hurych Z, Photoconductivity in crystalline and amorphous Sb_2O_3, Phys. Status Solidi (a), 16 (1973) K161-K163.
http://dx.doi.org/10.1002/pssa.2210160256

[31]   Tigau N, Ciupina V and Prodan G, The effect of substrate temperature on the optical properties of polycrystalline Sb_2O_3 thin films, J. Cryst. Growth, 277 (2005) 529-535.
http://dx.doi.org/10.1016/j.jcrysgro.2005.01.056

[32]   Validzic I.L, Abazovic N. D, Mitric M, Lalic M. V, Popovic Z. S and Vukajlovic F. R, Novel organo-colloidal synthesis, optical properties, and structural analysis of antimony sesquioxide nanoparticles, J. Nanopart. Res., 15 (2013) 1347.
http://dx.doi.org/10.1007/s11051-012-1347-x

[33]   Momma K and Izumi F, Commission on crystallographic computing IUCr Newsletter, No.7 (2006) 106-119.

CHAPTER 9

Charge Density of Al Doped Lanthanum Orthoferrites

R. Saravanan, G. Gowri

Research centre and Post Graduate Department of Physics, The Madura College, Madurai-625011, Tamil Nadu, India

saragow@gmail.com, gowrikanna01@gmail.com

Abstract

Bulk Al doped lanthanum orthoferrites ($La_{1-x}Al_xFeO_3$-LAFO) have been prepared using the solid state reaction method, for three different concentrations of Al (x=0, 0.40, 0.50). The prepared samples have been characterized by a powder X-ray diffractometer, scanning electron microscope, energy dispersive analysis using X-ray, UV-Visible spectrometer and vibration sample magnetometer respectively. The X- ray data of the samples have been used for resolving the structure and the refined structure factors were used for the study of the charge density distribution in the unit cell of the prepared samples, using the maximum entropy method. The magnetic behaviour of the Al doped samples has been found to enhance significantly as studied from the VSM measurements. Due to the enhancement in ferromagnetic behavior of aluminium doped lanthanum orthoferrite, the material can be considered as a capable candidate for magnetic memory device applications within the doping limit.

Keywords

XRD, MEM, Charge Density, VSM, UV-Visible Spectra

Contents

1. INTRODUCTION

In the current scenario, the simultaneous existence of magnetic (ferro/antiferro magnetic) and electric (ferro/antiferro electric) ordering in the rare earth orthoferrites (ABO_3) attract the researchers as one of the fascinating topic of research for their potential application in various electromagnetic devices. The advantage of ABO_3 perovskite structure is the replaceability of metallic ions at both A and B sites by various transition metals. In ABO_3 perovskite materials, the smaller cations B are at the centre of octahedron of oxygen anions and the large cations A are at the corners of the unit cell [1,2]. Many compounds having the structure similar to ABO_3 crystallize with the orthorhombic distortion in the perovskite structure and $LaFeO_3$ is the prototype of the series [2]. The crystal structure of $LaFeO_3$ is derived from the standard cubic structure through the distortion of the BO_6 octahedra [3-5]. Lanthanum orthoferrite is a significant material among the rare earth orthoferrites series, for its interesting properties viz., wide-gapped antiferromagnetic insulator with high Néel temperature ($T_N \sim 740°C$) [6], coexistence of coupled ferroelectric and antiferromagnetic ordering etc. [7]. Many remarkable observations on $LaFeO_3$ and doped $LaFeO_3$ have been reported which can advance the aspect of applications of lanthanum orthoferrite [6-13]. It has been studied that lanthanum orthoferrite is chemically stable in both reducing and oxidizing atmosphere [9]. In view of the fact that the doped $LaFeO_3$ has fascinating assorted properties like high electrical conductivity, excellent thermal stability, high dielectric constant, reasonable permittivity, ferroelectricity. It can be used as a separator material in solid oxide fuel cells [10-12], as hot electrode for magneto hydrodynamic power generation [13] as doped lanthanum ferric oxide shows oxygen ion conductivity. It is further used in oxygen permeable membrane [14-16], catalytic activity in the complex oxidation of hydrocarbons and catalytic combustion of methane [17]. Doped $LaFeO_3$ material is used as sensors and can be employed in microwave dielectrics and also used as solid electrolytes [18-20]. Doped $LaFeO_3$ is used as tunable capacitors and forms the source of ferroelectric random access memory for computers because of its ferroelectric nature. Due to its ferromagnetic nature, it is extensively used for recording and storing data, such as in hard drives [21]. Although a number of papers linked to the synthesis, structural, morphological, magnetic, dielectric and optical properties of pure and doped $LaFeO_3$ have been reported, no work has yet been reported concerning the charge density distribution in the unit cell of $LaFeO_3$ system. This has inspired the authors to synthesis $(La_{1-x}Al_xFeO_3)$ orthoferrites and to analyse the distribution charges in the unit cell and also the redistribution charges due to the variation in concentration of the dopant material Al that reveals the magnetic behaviour of Al doped $LaFeO_3$ system. In the present work, to compare the magnetic properties of pristine $LaFeO_3$ and Al doped $LaFeO_3$ orthoferrites, we prepared the samples using the solid state reaction method. The samples prepared in this work, have been characterized by X-ray Diffraction (XRD), scanning electron microscope (SEM), energy dispersive analysis using X-ray (EDAX), for structural, morphological, elemental compositional analysis. The UV-visible spectra were also taken into account for the prepared samples to study their optical properties. Finally, the magnetic parameters of the samples have been measured using a vibrating sample magnetometer (VSM). For the structural analysis, the Rietveld technique [22] has been employed on the XRD profile of the samples and hence the charge density distribution in the unit cell and also the bonding features have been investigated for the prepared orthoferrites. In the present work, it has been found that the Al substitution considerably enhances the magnetic properties in the doped samples as reported earlier [23] and the enhancement in magnetic behavior has been explained using charge density analysis.

2.   EXPERIMENTAL PROCEDURE

2.1   SAMPLE PREPARATION

Bulk pristine and aluminium doped lanthanum orthoferrites used in this work were prepared by the solid state reaction method using high purity oxides La_2O_3 (99.99 %, Alfa Aesar), Al_2O_3 (99.997 %, Alfa Aesar) and Fe_2O_3 (99.99 %, Alfa Aesar) as starting materials. These oxides in the desired stoichiometric proportions were weighted and then mixed thoroughly in an agate mortar to produce a uniform mixture. Circular pellets of 12 mm in diameter and 1 mm thickness were fabricated from the oxide mixtures and the pellets of these oxides mixture were heat treated in air at 1000 °C for 12 hours and then at 1100 °C for 14 hours and then finally, at 1300 °C for 16 hours with intermediate grinding to prepare the solid solutions of the samples.

2.2 RESULTS AND DISCUSSION

2.2.1 X-RAY PROFILE ANALYSIS

Bulk lanthanum orthoferrites $La_{1-x}Al_xFeO_3$ (x = 0, 0.40 and 0.50) were synthesized using the solid state reaction method and X-ray powder XRD data sets were obtained using an X-ray diffractometer, Bruker AXS D8 Advance, with CuK_α monochromatic beam at Sophisticated Analytical Instrument Facility (SAIF), Cochin university, Cochin, India, in the 2θ range of 5°-120° with a step size of 0.02° in 2θ. Figure 1 (a) shows the XRD patterns of the sintered samples. The first prominent (121) peak is enlarged and is shown in Figure 1 (b).

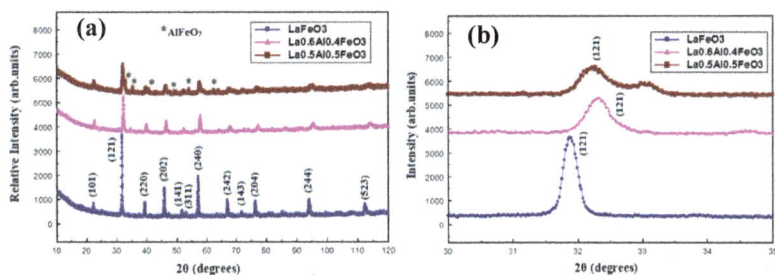

Figure 1. (a) Observed XRD pattern of $La_{1-x}Al_xFeO_3$ orthoferrites (b) Enlarged view of (121) peak

All the observed X-ray peaks for the prepared orthoferrites were identified and matched with the $LaFeO_3$ phase, standard pattern of the Joint Committee for Powder Diffraction Standards (JCPDS) XRD data set reported in the file (No.37-1493). The XRD results indicate that the products are a perovskite oxide with orthorhombic system with the space group of pnma. No additional phase has been identified for x=0 and x=0.40 which results that the integrated Al atoms are occupied in the preferential La lattice sites in the host lattice. In the high doping level of Al x=0.50, an additional phase of $AlFeO_3$ has been identified with $La_{0.5}Al_{0.5}FeO_3$ orthoferrite. The additional phase is compared and matched with JCPDS file (No.30-0024) and found to have an orthorhombic system of space group $pc2_1n$. The additional peaks corresponding to $AlFeO_3$ are shown in figure 1(a). It is observed from the XRD patterns that, for $La_{0.6}Al_{0.4}FeO_3$ orthoferrite, all the diffraction peaks corresponding to the hkl planes are shifted towards higher Bragg's angles from the diffraction peaks of pristine $LaFeO_3$. This confirms that the doping effect has occurred perfectly in the host lattice, up to x=0.4. But, due to the presence of $AlFeO_3$ an additional phase in $La_{0.5}Al_{0.5}FeO_3$ orthoferrite, all its diffraction peaks are shifted towards lower Bragg's angles from the diffraction peaks of pure $LaFeO_3$. It is also observed from figure

1, with the increasing of Al content, the intensity of each peak for the prepared sample tends to decreases. This trend also confirms the occurrence of a perfect doping effect. The reason for the reduction in the intensity is that the heavier atom La (Z= 57) present in the host lattice is replaced by the lighter atom Al (Z= 13).

To analyze the effect of Al substitution on the structural parameters of the prepared perovskites, the X-ray diffraction data of the prepared samples were subjected to the well-known Rietveld refinement technique [22], which is employed in the software JANA2006 [24]. The Rietveld refinement [25] is the standard tool which is devised by Hugo Rietveld [22] for the use of characterization of crystalline materials. The principle of the Rietveld [22] method is to minimize the difference between the theoretically modelled profile and the observed one.

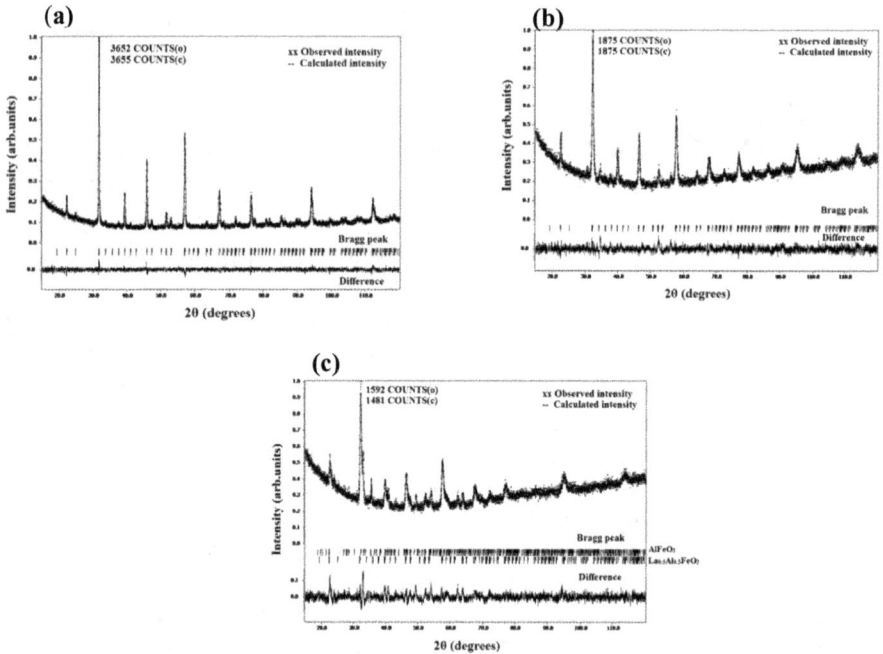

Figure 2. Rietveld refinement profiles of powder XRD data of (a) $LaFeO_3$ (b) $La_{0.6}Al_{0.4}FeO_3$ (c) $La_{0.5}Al_{0.5}FeO_3$

The refinement of the powder XRD data sets was done considering the orthorhombic structure for ($La_{1-x}Al_xFeO_3$) orthoferrites having a space group of pnma with 4 molecules per unit cell. The refined profile for the prepared orthoferrites is shown in figure 2 and the refined structural parameters are given in table 1.

Table 1. Structural parameters of $La_{1-x}Al_xFeO_3$ orthoferrites through refinement of powder XRD data

Parameter	$LaFeO_3$	$La_{0.6}Al_{0.4}FeO_3$	$La_{0.5}Al_{0.5}FeO_3$
a (Å)	5.5626(2)	5.5081(11)	5.4977(16)
b (Å)	7.8459(4)	7.7884(15)	7.7796(23)
c (Å)	5.5502(2)	5.5305(21)	5.5263(85)
$\alpha = \beta = \gamma$ (°)	90	90	90
Cell Volume, V (Å³)	242.2358(28)	237.2636(54)	236.3651(26)
Density, ρ (gm/ cc)	6.6540(3)	5.5406(33)	5.2472(80)
Observed reliability factor, Robs (%)	2.79	4.65	5.46
Profile reliability factor, Rp (/%)	4.17	3.95	4.44
GOF	1.05	1.14	1.36
Number of electrons in the unit cell	428	358	340

The refined positional coordinates are given in table 2. From the structural refinements it is observed that the unit cell value, unit cell volume and also unit cell density decreases with the increase of Al content as given in table 1, which is due to the lower ionic radius of Al^{3+} (r_{Al} = 0.535 Å) than that of La^{3+} (r_{La} = 1.032 Å) [26].

Table 2. Refined atomic positional coordinates of $La_{1-x}Al_xFeO_3$ orthoferrites

Atom	$LaFeO_3$			$La_{0.6}Al_{0.4}FeO_3$			$La_{0.5}Al_{0.5}FeO_3$		
	x	y	z	x	y	z	x	y	z
La/Al	0.0294	0.25	0.9926	0.0244	0.25	0.9923	0.0182	0.25	0.9937
Fe	0.5	0	0	0.5	0	0	0.5	0	0
O_1	0.4867	0.25	0.0699	0.488	0.25	0.0707	0.480	0.25	0.0707
O_2	0.2831	0.0387	0.7168	0.2642	0.0436	0.8379	0.2831	.0387	0.7168

2.2.2 CHARGE DENSITY ANALYSIS

PXRD data sets of the prepared samples were refined using the JANA2006 software [24]. Then, for the chosen systems, the charge density distribution in the unit cell has been determined via the maximum entropy method (MEM) [27] using the PRIMA software (Practice of Iterative MEM Analysis) [28] and the results are visualized using the visualization software VESTA (Visualization for Electronic and STructural Analysis) [29]. MEM is a precise tool to study the electron density distribution due to its resolution. Using this technique, the bonding nature and the distribution of electrons in the bonding region can be clearly visualized. MEM electron densities are always positive and even with limited number of data, reliable electron densities approaching true densities can be estimated. The MEM refinements in this work were carried out by dividing the unit cell into 48 x 72 x 48 pixels. The MEM refined parameters are given in Table 3.

Table 3. Parameters from MEM refinements of $La_{1-x}Al_xFeO_3$ orthoferrites

Parameter	$LaFeO_3$	$La_{0.6}Al_{0.4}FeO_3$	$La_{0.5}Al_{0.5}FeO_3$
Number of pixels in the unit cell (48x72x48)	165888	165888	165888
Number of electrons in the unit cell	428	358	340
Lagrangian parameter, λ	0.011985	0.011700	0.012220
Number of refinement cycles	1398	1341	1445
Reliability factor, R_{MEM} (%)	0.013361	0.020395	0.027840
Weighted reliability factor, wR_{MEM} (%)	0.016776	0.021646	0.026547

The three dimensional picture of charge density in the unit cell with isosurface level, 5.5 $e/Å^3$, is drawn for the prepared orthoferrites and is presented in figure 3. The site of La, Fe and O atoms are clearly visualized through the charge distribution. The shaded region enclosed by the electron cloud shows an atom.

The two dimensional electron density distributions in the (100) miller plane for Fe and O1 atoms are shown in figures 4(a), 4(b), 4(c), for x=0, 0.4 and 0.5 respectively. Similarly, figures 5(a), 5(b), 5(c) show the two dimensional electron density distributions in the (200) miller plane for La and O1 atoms, for x=0, 0.4 and 0.5 respectively. From the 2D electron density maps, it is evident that for the doped systems, the contour lines are elongated from Fe to O1 atom and also from La to O1 atom, which is an indication of the decrease in the ionic behavior of both the bonds in the doped systems. However, in the

$La_{0.5}Al_{0.5}FeO_3$ system the elongation of contour lines are observed to be less due to the presence of an additional phase $AlFeO_3$, which indicates that there is a slight increase in the ionic behavior of both the bonds compared to that of the $La_{0.6}Al_{0.4}FeO_3$ system.

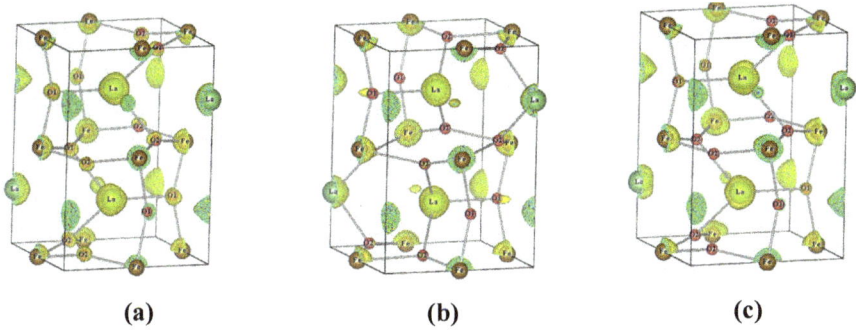

(a)   (b)   (c)

Figure 3.   Three dimensional electron density distribution in the unit cell (isosurface level = 5.5 e/ Å^3) for (a) LaFeO$_3$ (b) La$_{0.6}$Al$_{0.4}$FeO$_3$ (c) La$_{0.5}$Al$_{0.5}$FeO$_3$

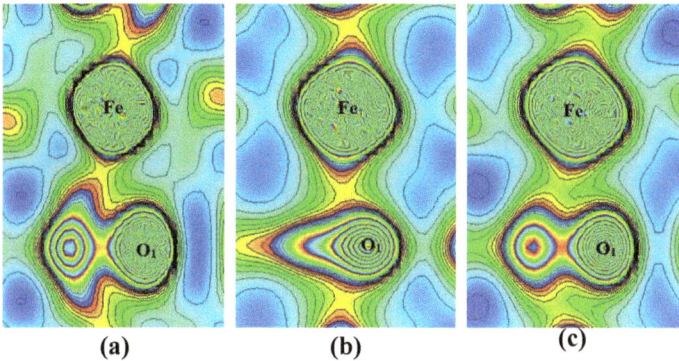

(a)   (b)   (c)

Figure 4.   Two dimensional electron density distribution in the unit cell showing Fe–O$_1$ bond in the (1 0 0) plane in the contour range of 0 to 1.4 e/ Å^3 with contour interval of 0.1 e/ Å^3 for (a) LaFeO$_3$ (b) La$_{0.6}$Al $_{0.4}$FeO$_3$ (c) La$_{0.5}$Al$_{0.5}$FeO$_3$

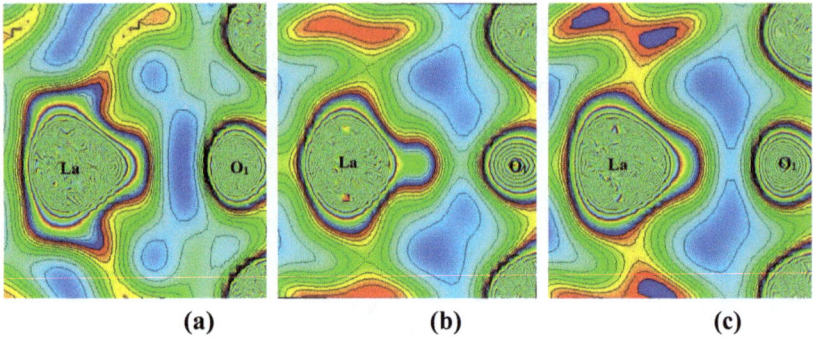

(a)   (b)   (c)

Figure 5.   Two dimensional electron density distribution in the unit cell showing La–O_1 bond in the (2 0 0) plane in the contour range of 0 to 1.4 e/ $Å^3$ with contour interval of 0.1 e/ $Å^3$ for (a) $LaFeO_3$ (b) $La_{0.6}Al_{0.4}FeO_3$ (c) $La_{0.5}Al_{0.5}FeO_3$

To analyse the electronic distribution quantitatively, the one dimensional profile of the electron density between the Fe and O_1 atoms and also for the La and O_1 atoms is drawn as shown in Figure 6. For the prepared samples, the value of the charge density at the bond critical point along the Fe–O_1 bond and the La–O_1 bond is estimated and is given in table 4.

(a)   (b)

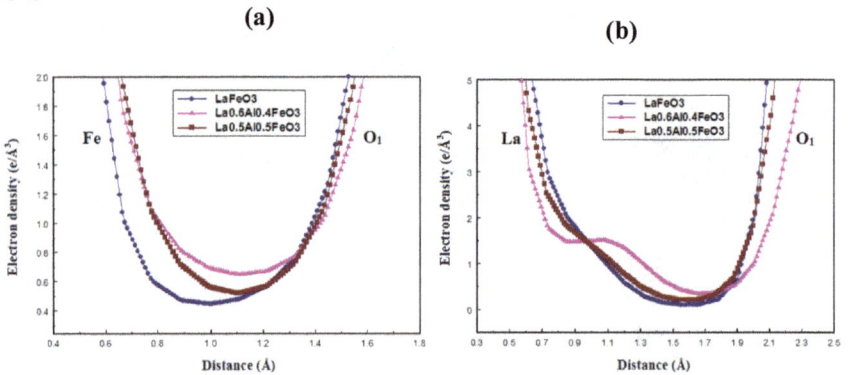

Figure 6.   One dimensional electron density profiles along (a) Fe–O_1 bond, (b) La–O_1 bond

*Table 4.   Charge density at the bond critical point between atoms in $La_{1-x}Al_xFeO_3$
orthoferrites*

Sample	Bond	Distance (Å)	Charge density $(e/Å^3)$	Bond length (Å)
$LaFeO_3$	Fe-O_1	0.9954	0.4462	2.0008
	La-O_1	1.5570	0.1092	2.4398
$La_{0.6}Al_{0.4}FeO_3$	Fe-O_1	1.1083	0.6487	1.9870
	La-O_1	1.6695	0.3493	2.4251
$La_{0.5}Al_{0.5}FeO_3$	Fe-O_1	1.1071	0.5206	1.9848
	La-O_1	1.5399	0.2173	2.4129

The value of the charge density at the BCP clearly indicates that at x=0, the bonding
between Fe and O_1 atoms and La and O_1 atoms show an electron density with the values
as 0.4462 e/$Å^3$ (at Fe-O_1 bond) and 0.1092 e/$Å^3$ (at La-O_1 bond). At x=0.4, the value of
the charge density at BCP along the Fe–O_1 bond and also along the La–O_1 bond, further
increases showing higher values as 0.6487 e/$Å^3$ (at Fe–O_1 bond) and 0.3493 e/$Å^3$ (at La–
O_1 bond) which leads to decrease of the ionic behaviour of the system and it results in
ferromagnetic behaviour due to doping of diamagnetic ion Al^{3+} in the host lattice.
Further, increasing Al content as x=0.5 at the host lattice, gives the value of the charge
density at the BCP as 0.5206 e/$Å^3$ (at Fe–O_1 bond) and 0.2173 e/$Å^3$ (at La–O_1 bond)
showing a slight increase in the ionic behaviour of the system, because at x=0.5, the
system has an additional phase of $AlFeO_3$. Due to the presence of $AlFeO_3$ as additional
phase exhibiting antiferromagnetic behaviour [30], the electron orientation is
redistributed and it achieves the magnetic degradation from ferromagnetism to weak
ferromagnetism with less magnetization. Moreover, in the prepared orthoferrites, the
bond length is found to decreases with increase in Al dopant concentration. The reason
for this is due to the replacement of higher ionic radius La^{3+} (r_{La} = 1.032 Å) [26] in the
host lattice site by lower ionic radius of Al^{3+} (r_{Al} = 0.535 Å) [26], where the radii given
corresponds to the coordination number six, of La and Al atoms in the doped $LaFeO_3$
system.

2.2.3 SIZE AND ELEMENTAL COMPOSITIONAL ANALYSIS

To study the surface morphology and particle size of the prepared $(La_{1-x}Al_xFeO_3)$
orthoferrites, SEM micrographs have been recorded. The scanning electron microscope

(SEM) characterization has been done for the samples under investigation with different magnification. Figures 7(a_1 to c_1) show the SEM images for $La_{1-x}Al_xFeO_3$ (x=0, 0.40 and 0.50) orthoferrites with the magnification of 10,000. In the SEM images, it can be seen that the particles are almost irregularly deagglomerated. The average particle size of the prepared orthoferrites is determined from the SEM image and their values are listed in table 5.

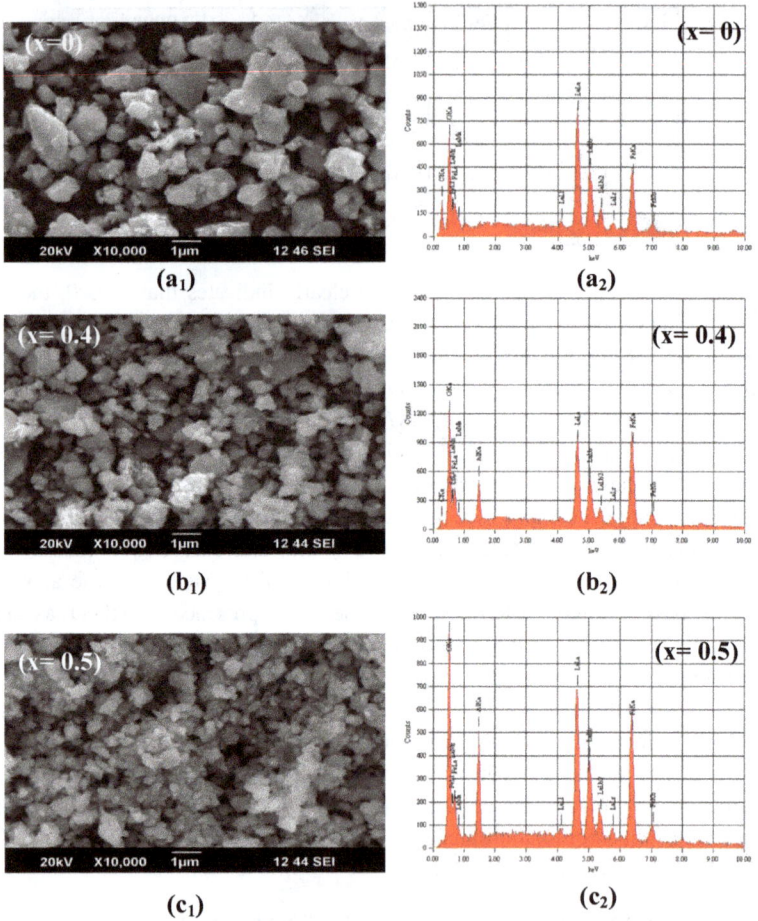

Figure 7. Morphological images (SEM) (a_1-c_1) and EDAX spectra (a_2-c_2) of $La_{1-x}Al_xFeO_3$ orthoferrites

The crystallite sizes of the prepared samples were also determined with the data, full width at half maximum (FWHM) of the powder X-ray peaks and its corresponding Braggs angles, using the Grain [31] software which employs the Debye Scherrer's formula $D = k\lambda/d \cos\theta$, where D is the crystallite size of the coherently diffracting domain which is different from the particle size, k is a constant usually of 0.9, λ the wavelength of X-rays (1.54056 Å) for CuK_α radiation, β is the full width at half maximum in radian for the prominent intensity peak, and θ is the Bragg angle of the reflection. The powder XRD gives only the size of the coherently diffracting domain and it is known as crystallite size. The calculated crystallite size of the prepared orthoferrites is listed in table 5. It is found that for the prepared samples, the particle size and crystallite size decreases with increase in Al content.

Table 5.   Crystallite size and particle size of $La_{1-x}Al_xFeO_3$ orthoferrites

Parameter	$LaFeO_3$	$La_{0.6}Al_{0.4}FeO_3$	$La_{0.5}Al_{0.5}FeO_3$
Crystallite size (nm)	42	33	4
Particle size (nm	711	578	343

The spectra from the energy dispersive analysis of X-rays show the elemental composition of materials in a quantitative way. The observed EDAX spectra for the prepared $La_{1-x}Al_xFeO_3$ orthoferrites are shown in figures 7(a_2 to c_2). The peaks in the spectra indicate which atomic species is present. The EDAX results show the main peaks of La, Al, Fe, and O atoms in all samples beside with peak of C atom which is used in the scanning electron microscope instrument as background material. These EDAX spectra confirm that the prepared lanthanum orthoferrites are free from impurities. The values of each element present in the prepared orthoferrites are tabulated and shown in table 6(a_3 to c_3).

2.2.4 UV ANALYSIS

To understand the optical property and to estimate the optical band gap energy of the prepared ($La_{1-x}Al_xFeO_3$) orthoferrites, UV-visible absorption spectral analysis has been done. UV-visible spectra were characterized at SAIF, Cochin, India, using a UV-visible – NIR spectrometer, Varian Cary 5000. In the UV-visible optical absorption spectrum of the prepared $La_{1-x}Al_xFeO_3$ (x=0, 0.40 and 0.50) orthoferrites, the sharp absorption peaks are identified in the UV-region at the wavelength 259 nm, 276 nm, 282 nm respectively.

In the absorption spectrum of the prepared samples, the infinitesimal red shift has been observed for Al doped samples, ensuing decrease in the optical energy band gap value.

The optical band gap for the prepared ($La_{1-x}Al_xFeO_3$) orthoferrites were determined according to the equation $(\alpha hv)=k (hv-Eg)1/n$ [32], where hv is photon energy (E), α is absorption coefficient, K is a material constant and n is 2 as $LaFeO3$ system is a direct band gap material. To estimate the band gap energy for the prepared ($La_{1-x}Al_xFeO_3$) orthoferrites, a graph is plotted for $(\alpha E)2$ versus energy as shown in figure 8.

Table 6. The elemental compositions (a_3-c_3) of $La_{1-x}Al_xFeO_3$ orthoferrites (a_3) x=0, (b_3) x=0.4, (c_3) x=0.5

Element	Mass%	Atom%
C K	7.42	31.39
O K	8.81	28
Fe K	18.28	16.64
La L	65.49	23.97
Total	100	100

(a_3)

Element	Mass%	Atom%
C K	1.59	7.09
O K	**10.88**	**36.53**
Al K	2.9	5.78
Fe K	31.1	29.91
La L	53.53	20.69
Total	100	100

(b_3)

Element	Mass%	Atom%
O K	12.28	42.2
Al K	4.66	9.49
Fe K	26.22	25.81
La L	56.84	22.5
Total	100	100

(c_3)

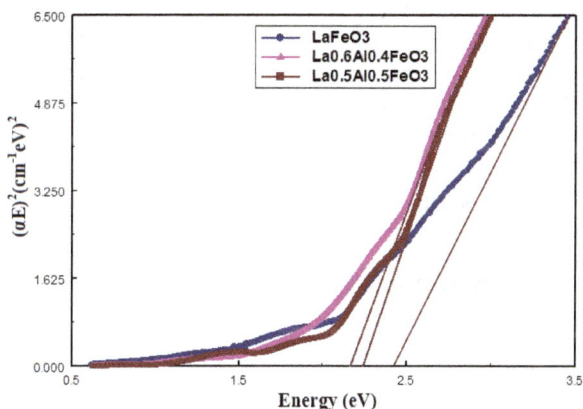

Figure 8. Plot of energy versus $(\alpha E)^2$ in finding the band gap for $La_{1-x}Al_xFeO_3$ orthoferrites

Then, the energy of the band gap is estimated by extrapolating the slope to the zero value of $(\alpha E)^2$. It has already been reported that the optical band gap of $LaFeO_3$ ceramics were found to be vary from 2.12 eV to 2.51 eV [33]. The estimated optical energy band gap values of the prepared samples are very close to the already reported values and are listed in table 7.

Table 7. UV-Visible optical parameters of $La_{1-x}Al_xFeO_3$ orthoferrites

Parameter	$LaFeO_3$	$La_{0.6}Al_{0.4}FeO_3$	$La_{0.5}Al_{0.5}FeO_3$
Wavelength (nm)	259.00	276.00	282.00
Optical absorption	0.920	0.879	0.885
Optical band gap Energy (eV)	2.4335	2.1681	2.2447

The energy band gap values are found to be decreased for the doped samples. However, the value of energy band gap for $La_{0.5}Al_{0.5}FeO_3$ system increases compared to that of the $La_{0.6}Al_{0.4}FeO_3$ system. The reason for decrease in energy band gap for the doped samples is analysed from the charge density distribution. From the charge density analysis, the value of electron density at the bond critical point between Fe and O_1 atoms and also La and O_1 atoms for Al doped LAFO system is found to be increased compared to that of pure $LaFeO_3$.

Thus the value of bond critical point charge density indicates that when the rare earth ions La^{3+} are replaced by semiconductor ions Al^{3+} in the $LaFeO_3$ system , the number of free electrons increases and they move from the valence band to the conduction band, O $2p{\rightarrow}Fe$ 3d. Due to the occupation of higher number of free electrons in the conduction band its band width increases and hence the gap between the valence band and the conduction band decreases. Consequently, this results in decrease in the value of energy band gap. Thus the conductivity of $LaFeO_3$ is enhanced due to the doping of Al in that system [23].

2.2.5 MAGNETIC BEHAVIOUR ANALYSIS

To analyze the magnetic behaviour of the prepared $(La_{1-x}Al_xFeO_3)$ orthoferrites, they are subjected to the magnetic field and the magnetic hysteresis loops have been recorded at room temperature and hence the magnetic parameters are also measured using a vibration sample magnetometer (Lakeshore VSM) at Sophisticated Analytical Instrument Facility (SAIF), Indian Institute of Technology of Madras, Chennai, India.

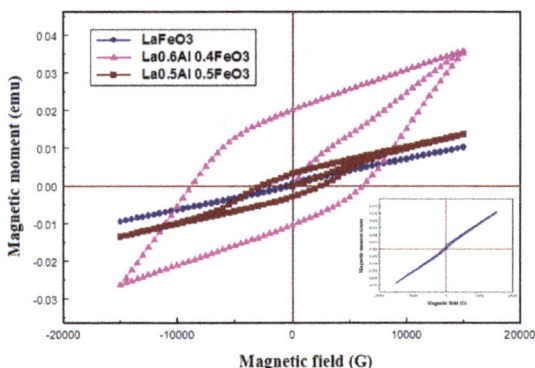

Figure 9.   Magnetic hysteresis (M vs H) curve of $La_{1-x}Al_xFeO_3$ orthoferrites. Inset shows the (M vs H) curve $LaFeO_3$ orthoferrite

The magnetic properties of the sample are usually characterized by its magnetic hysteresis loop that gives the magnetic behaviour of the prepared sample when subjected in an external magnetic field. Figure 9 shows the magnetic hysteresis loop of the prepared samples. From the magnetic hysteresis loop of the samples, it is observed that for a pure $LaFeO_3$ system, the loop of magnetization curve is very narrow and is shown in the inset of figure 9, indicating its weak ferromagnetic behaviour, while at x=0.4, that is for 40 %

of Al doped samples, the magnetic hysteresis curve show larger loop indicating the ferromagnetic behaviour with enhanced magnetization. However, at x=0.5, that is for 50% of Al doped samples, the loop opening is smaller when compared to 40 % of Al doped sample due to the presence of $AlFeO_3$ exhibiting antiferromagnetic bahaviour that reduces magnetization. Thus the magnetic hysteresis loop of the prepared lanthanum orthoferrites confirms the presence of magnetic ordering in the samples. The values of magnetic parameters extracted from the hysteresis loop are listed in table 8.

Table 8. Magnetic parameters of $La_{1-x}Al_xFeO_3$ orthoferrites obtained from VSM

Parameter	$LaFeO_3$	$La_{0.6}Al_{0.4}FeO_3$	$La_{0.5}Al_{0.5}FeO_3$
Magnetization, $M_s \cdot 10^{-3}$ (emu)	9.9146	31.059	13.568
Coercivity, H_c (G)	316.17	7450.7	2608.9
Retentivity, $M_r \cdot 10^{-3}$ (emu)	0.3461	15.254	3.0627

The reason for the enhanced magnetic behaviour can be correlated with the electron density distribution. From the charge density distribution it is observed that the value of charge density at the bond critical point between Fe and O_1 atoms and also La and O_1 atoms increases for the doped samples. In general, below Néel temperature, $LaFeO_3$ is a wide-gapped antiferromagnetic insulator [6], where their magnetic moments are due to the ordering of the uncompensated spin of Fe^{3+} ions while the magnetic moment of La^{3+} ions are paramagnetic. At Néel temperature, antisymmetric exchange interaction takes place between the La^{3+} and the Fe^{3+} ions that makes the La^{3+} ions to acquire magnetization, resulting the $LaFeO_3$ system to be a weak ferromagnetic one. For this system, the values of coercivity, saturation magnetization and remanent magnetization are 316.17 G, 0.0099 emu and 0.0003 emu respectively. In addition to this, 40 % doping of diamagnetic Al^{3+} ions [34] with this system, reduces the grain size that leads to increase in the number of ferromagnetic domains. Consequently, this results in the increase in the number of electrons with uncompensated spin in the d shell of Fe^{3+} ions that are responsible for the increase in the magnetic moments. For $La_{0.6}Al_{0.4}FeO_3$ system, the values of coercivity, saturation magnetization and remanent magnetization are increased as 7450.7 G, 0.031 emu and 0.0152 emu respectively. However in the 50 % doped samples, though the doping reduces grain size which leads to the increase in the number of ferromagnetic domains, the presence of antiferromagnetic material $AlFeO_3$ decreases the number of effective ferromagnetic domain. This results in the reduction of the number of electrons with uncompensated spin in the d shell of Fe^{3+} ions and also in the resultant magnetic moments. Thus for $La_{0.5}Al_{0.5}FeO_3$ system, instead of increasing the magnetic parameter values than that of 40% doped sample, the values of coercivity, saturation magnetization and remanent magnetization are decreased as 2608.9 G, 0.0135 emu and 0.003 emu respectively. Since, $La_{0.6}Al_{0.4}FeO_3$ orthoferrite has a high coercivity and also has a high saturation magnetization and remanent magnetization that are desirable for permanent magnets; it can be used as a promising candidate in magnetic recording and memory devices.

3 CONCLUSION

$La_{1-x}Al_xFeO_3$ (x= 0, 0.4, 0.5) orthoferrites were successfully synthesized using the solid state reaction method. XRD analysis reveals that prepared samples exhibit an orthorhombic phase of space group pnma. The additional phase of $AlFeO_3$ is found in the doped sample at x=0.5. SEM images show that particle size reduces as the Al content increases. The EDAX result clearly shows only the main peaks of La, Al, Fe, and O atoms in all samples. The UV-visible spectra of the samples show that Al doping decreases the optical energy band gap values significantly. The magnetic parameters extracted from the VSM measurement show that the substitution of diamagnetic ions Al^{3+} in the $LaFeO_3$ system enhances the ferromagnetic behaviour at x=0.4, whereas the system loses its ferromagnetic behaviour at x=0.5, due to the presence of $AlFeO_3$ as additional phase exhibiting antiferromagnetic bahaviour. The electronic distribution in the unit cell is analyzed through the MEM method for the prepared lanthanum orthoferrites. The bonding features of the prepared samples are also analyzed. The change in magnetic behaviour with respect to Al doping are correlated with the charge density distribution in the unit cell. The doping limit of Al in the $LaFeO_3$ system is found to be within x=0.5.

REFERENCES

[1]   J.-M. Liu, Q.C. Li, X.S. Gao, Y. Yang, X.H. Zhou, X.Y. Chen, et al.,Order coupling in ferroelectromagnets as simulated by a Monte Carlo method, Phys.Rev.B. 66 (2002) 054416–054426.
       http://dx.doi.org/10.1103/PhysRevB.66.054416

[2]   N.A. Hill, Why are there so few magnetic ferroelectrics? J.Phys. Chem. B.104 (2000) 6694–6709.
       http://dx.doi.org/10.1021/jp000114x

[3]   Y.-H. Lee, J.-M. Wu, Epitaxial growth of LaFeO3 thin films, J. Cryst. Growth. 263 (2004) 436-441.
       http://dx.doi.org/10.1016/j.jcrysgro.2003.12.007

[4]   P.M. Woodward, Octahedral Tilting in Perovskites. I. Geometrical Considerations, Acta Cryst. B53 (1997) 32-43.
        http://dx.doi.org/10.1107/S0108768196010713

[5]   P.M. Woodward, Octahedral Tilting in Perovskites. I. Geometrical Structure, Acta Cryst. B 53 (1997) 44-66.
        http://dx.doi.org/10.1107/S0108768196012050

[6]   G.R. Hearne, M.P. Pasternak, Electronic structure and magnetic properties of LaFeO3 at high pressure, Phys. Rev. B. 51 (1995) 11495–11500.
        http://dx.doi.org/10.1103/PhysRevB.51.11495

[7]   S. Acharya, J. Mondal, S. Ghosh, S.K. Roy, P.K. Chakrabarti, Multiferroic behavior of Lanthanum orthoferrite(LaFeO3), Mater. Lett. 64 (2010) 415-418.
        http://dx.doi.org/10.1016/j.matlet.2009.11.037

[8]   M.A. Ahmed, S.I. El-Dek, Extraordinary role of Ca 2+ ions on the magnetization of LaFeO3 orthoferrite, Mater. Sci. Eng. B, 128 (2006) 30–33.
        http://dx.doi.org/10.1016/j.mseb.2005.11.013

[9]   S. Farhadi, Z. Momeni, M. Taherimehr, Rapid synthesis of perovskite-type LaFeO3 nano particles by microwave- assisted decomposition of bimetallic [LaFe(CN)6].5H2O Compound, J. Alloys Compd. 471 (2009) L5–8.
        http://dx.doi.org/10.1016/j.jallcom.2008.03.113

[10]   F. Li, Y. Liu, R. Liu, Z. Sun, D. Zhao, C. Kou, Ferromagnetism and optical properties of La1 − x Al x FeO3 nanopowders, Mater. Lett. 64 (2010) 223–225.
        http://dx.doi.org/10.1016/j.matlet.2009.10.048

[11]   A. Singh, R. Chatterjee, Magnetization induced dielectric anomaly in multiferroic solid solution, Appl. Phys. Lett. 93 (2008) 182908–182910.
        http://dx.doi.org/10.1063/1.3012389

[12]   J. Lüning, F. Nolting, A. Scholl, H. Ohldag, J. W. Seo, J. Fompeyrine, J.-P. Locquet, and J. Stöhr, et al., Determination of the antiferromagnetic spin axis in epitaxial films by x-ray magnetic linear dichroism spectroscopy, Phys. Rev. B. 67 (2003) 214433–214436.
        http://dx.doi.org/10.1103/PhysRevB.67.214433

[13]   J.W. Seo, E.E. Fullerton, F. Nolting, A. Scholl, J. Fompeyrine, J.P. Locquet, Exchange Bias Induced by the Fe3O4 Verwey transition, J. Phys.: Condens. Matter. 20 (2008) 264014–264023.
        http://dx.doi.org/10.1088/0953-8984/20/26/264014

[14]   S.J.Chao, K.S. Song, I.S. Ryu, Y.S. Seo, M.W. Rayoo, S.K.Kang, Catal. Lett. 58 (1999) 63–66.
       http://dx.doi.org/10.1023/A:1019092809562

[15]   N.Q. Minh, Ceramic fuel cells, J. Am. Ceram. Soc. 76 (1993) 563–588.
       http://dx.doi.org/10.1111/j.1151-2916.1993.tb03645.x

[16]   B.C.H. Steele, Studies on electrical and dielectric properties of LaFeO3, Mater. Sci. Eng.B. 13 (1992) 75–79.
       http://dx.doi.org/10.1016/0921-5107(92)90146-Z

[17]   D.B. Meadowcraft J.M. Wimmer, Studies on electrical and dielectric properties of LaFeO3, Ceram. Bull. 58 (1979) 610–612.

[18]   G. Karlsson, Studies on electrical and dielectric properties of LaFeO3, Electrochim. Acta. 30 (1985) 1555–1561.
       http://dx.doi.org/10.1016/0013-4686(85)80019-0

[19]   T.Nekamura, G.Petzow, L.J.Gauckler, Stability of the perovskite phase LaBO3 (B=V, Cr, Mn, Fe, Co, Ni) in reducing atmosphere. I. Experimental results, Mater. Res. Bull. 14 (1979) 649–659.
       http://dx.doi.org/10.1016/0025-5408(79)90048-5

[20]   J.Mizusaki, T. Sisamoto, W.K.Cannon, H.Kent Bowen, Studies on electrical and dielectric properties of LaFeO3, J. Am. Ceram. Soc. 65 (1982) 363.
       http://dx.doi.org/10.1111/j.1151-2916.1982.tb10485.x

[21]   M.A.Ahmed, N.Okasha, B. Hussein, Synthesis, characterization and studies on magnetic and electrical properties of LaAlyFe1-yO3 nanomultiferroic, J. Alloys Compd. 553 (2013) 308–315.
       http://dx.doi.org/10.1016/j.jallcom.2012.11.114

[22]   H.M. Rietveld: A Profile Refinement Method for Nuclear and Magnetic Structures J. Appl. Crystallogr. 2 (1969) 65-71.
       http://dx.doi.org/10.1107/S0021889869006558

[23]   S. Acharya, P.K. Chakrabarti, Some interesting observations on the magnetic and electric properties of Al3+ doped lanthanum orthoferrite (La0.5Al0.5FeO3), Solid State Commun. 150 (2010) 1234-1237.
       http://dx.doi.org/10.1016/j.ssc.2010.04.006

[24]   Petříček, V., Dušek, & M., Palatinus, L. JANA 2006, the crystallographic computing system. Praha, Czech Republic: Academy of Sciences of the Czech Republic, (2006).

[25]   M. M. Wolfson. Introduction to X-ray Crystallography, Cambridge University Press, London, 1970.

[26]   R.D.Shannon, Revised Effective Ionic Radii and Systematic Studies of Interatomic Distances in Halides and Chalcogenides. Acta Cryst. A32 (1976) 751-767.
http://dx.doi.org/10.1107/S0567739476001551

[27]   M. Sakata, M.Sato, Accurate structure analysis by the maximum entropy method, Acta crystallographica. A46 (1990) 263-270
http://dx.doi.org/10.1107/S0108767389012377

[28]   F.Izumi, R.A.Dilanien. PRIMA, for the maximum entropy method advanced materials laboratory, Japan, 2004.

[29]   K.Momma and F.Izumi, VESTA: a three-dimensional visualization system for electronic and structural analysis, J.Appl. Crystallogr, 41 (2008) 653-658.
http://dx.doi.org/10.1107/S0021889808012016

[30]   R.Caracas, Spin and structural transitions in AlFeO3 and FeAlO3 perovskite and post-perovskite, Phys. Earth Planet. In. 182 (2010) 10-17.
http://dx.doi.org/10.1016/j.pepi.2010.06.001

[31]   R. Saravanan, GRAIN software, Private Communication. 2008.

[32]   S.M. Sze, Physics of Semiconductor Devices, J. Wiley & Sons, 1969, 52.

[33]   Roberto Köferstein, Lothar Jäger, Stefan G. Ebbinghaus, magnetic and optical investigations on LaFeO3 powders with different particle sizes and corresponding ceramics, Solid State Ionics, (2013) 249-250:1-5.
http://dx.doi.org/10.1016/j.ssi.2013.07.001

[34]   M.A. Ahmed, N.Okasha, B.Hussein, Enhancement of the magnetic properties of Al/La multiferroic, J. Magn. Magn. Mater. 324 (2012) 2349-2354.
http://dx.doi.org/10.1016/j.jmmm.2012.02.036

CHAPTER 10

Charge Density Distribution and Bonding in Calcite

T. K. Thirumalaisamy[1], S. Saravanakumar[2], R. Saravanan[3],

[1]Department of Physics, H.K.R.H. College, Uthamapalayam-625 533, Tamil Nadu, India.

[2]Department of Physics, Kalasalingam University, Krishnankoil – 626 126, Tamil Nadu, India

[3]Research Centre and PG Department of Physics, The Madura College, Madurai-625 011, Tamil Nadu, India

Email: tktsamy67@gmail.com; saravanaphysics@gmail.com; saragow@gmail.com

Abstract

An attempt to characterize the bonding and visualization of charge density distribution in calcite is achieved. From the X-ray diffraction data sets the experimental charge density distribution and its derived properties in calcite are derived and analyzed using an aspherical atom based multipole model refinement and the maximum entropy method (MEM). The multipole analysis is done for the refinement of the population parameters. The topology of the charge density is analyzed and the critical points in the charge density are determined. The covalent nature of the bonding between C – O is revealed in the 3D, 2D MEM maps and also in the one-dimensional electron density profiles. The quantitative analysis of the bonding is done using the charge density profiles along the bond path. The density at bond critical point along the bonding direction is found to be around 1.7856 e/Å^3 and 0. 0.4994 e/Å^3 for C –O and Ca – O respectively.

Keywords

Charge Density, Planar CO_3, Multipole, MEM.

Contents

1. INTRODUCTION

Calcite is one of the most important carbonate minerals on the surface of the earth. The shapes and forms of calcium carbonate encountered in nature strongly contrast those that are generally formed in a synthetic environment. Many researchers have engaged in unravelling the mysteries of calcium carbonate biomineralization, either by analyzing biogenic materials or by trying to understand the interactions between organic and inorganic phases in a laboratory environment. Recently, scientists have been trying to communicate the crystal habit with polymorph in the crystals. Calcium carbonate occurs in nature as three polymorphs (calcite, aragonite and vaterite) in two hydrated crystal forms (calcium carbonate monohydrate and calcium carbonate hexahydrate) and also as amorphous material [4-6]. The most stable calcite, less stable aragonite and even less stable vaterite have a rhombohedral, orthorhombic and hexagonal crystallographic unit cell [7, 8] respectively. The morphology of the crystal is predicted by the Wulff's [9] rule, which states that the most stable habit is related to the minimum of the sum of the products between crystal surface area and surface energy for the different faces of the crystal. According to Bridgman [10], calcite had undergone phase transitions at a pressure of 1.44 GPa and 1.77 GPa to slightly denser phase, calcite II and significantly denser phase, calcite III respectively. Singh and Kennedy [11] have identified the calcite I - II transformation at 1.45 GPa and the calcite II – III transformation at 1.74 GPa at room temperature. Merrill and Bassett [12] have noticed the calcite I - II transition at 1.5 GPa and the II - III transition at 2.2 GPa.

In the present analysis, X-ray powder diffraction (XRPD) data set of calcite has been refined using the Rietveld [13] method and then the maximum entropy method (MEM) [14] has been applied to illustrate the charge density distribution in calcite. The size of the particles evaluated using scanning electron microscope (SEM) and X-ray powder diffraction (XRPD) have been reported. The band gap energy has also been estimated by using UV-visible analysis. An attempt to the new understanding of the bonding and charge density distribution of the older non-linear optical material calcite has been done. And so it was our curiosity to understand the adoptability of the models for the present study which lead us to the estimation and analysis of the charge density in calcite using both the multipole method and maximum entropy method (MEM) [14].

2. EXPERIMENTAL

The calcite powder was prepared from single crystal calcite material. The single crystal calcite has been purchased from scientific suppliers. The mechanical milling was applied in the single crystal calcite material using an agate mortar pistol for 5 h. After grinding, the powdered calcite was sieved. The prepared sample was characterized by an X-ray diffractometer (XRD). The X-ray intensity data was collected at Sophisticated Analytical Instruments Facility (SAIF), Department of Science and Technology (DST), Cochin, India, using a parallel-beam Bruker AXS D8 Advance (Karlsruhe, Germany), X-ray diffractometer fitted with Si (Li) detector type. Soller silt set to a 6° (2θ) aperture was used to improve the peak shape and theta/2theta geometry. The accelerating voltage and the applied current density were 40 KV and 35 mA/cm^2 respectively. The wavelength used for the X-ray intensity data collection was 1.5418 Å with a 2θ range of data collection from 5° to 120° with a step size of 0.01° and counting time of 49.2 s at each step.

Scanning electron microscope (SEM) micrographs of calcite was obtained under different magnifications on a field emission SEM apparatus (JSM-6390LV, JEOL) operating at acceleration voltage of 30 kV, one of which is shown in figure 1. The particle size of calcite from SEM measurement is ~0.4 μm.

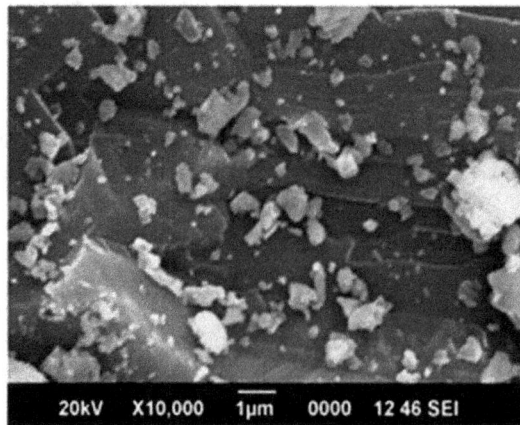

Figure 1. SEM micrograph of calcite.

The optical band gap energy (E_g) was evaluated according to the equation proposed by Wood et al. [15]. The band gap of the calcite was estimated by plotting $(\alpha h\nu)^2$ versus hν and extrapolating the linear portion near the onset of absorption edge to the energy axis.

Figure 2 shows the UV-visible spectrum of calcite. In this work, the value of the band gap energy of calcite is found to be 3.90 eV and it agrees well with the earlier reported value of 3.93 eV [16].

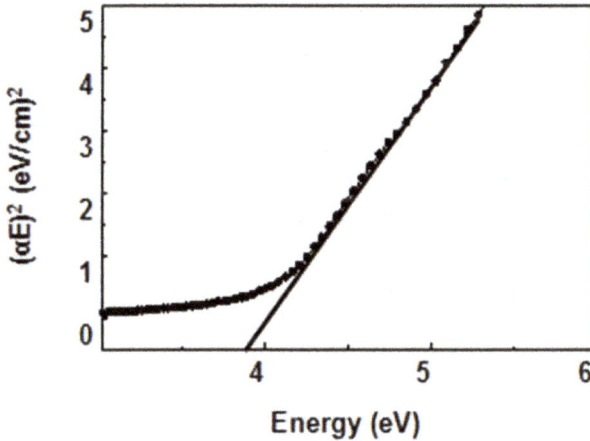

Figure 2. UV-vis spectrum of calcite.

3. METHODOLOGY

The raw intensities were refined based on the Rietveld [13] method using the software program JANA2006 [17]. The Rietveld [13] method refines structural parameters like fractional co-ordinates, atomic displacement parameters, occupation factors and lattice parameters from the whole powder diffraction patterns. In this method, the observed profiles are matched with the profiles constructed similarly using the pseudo-Voigt profile shape function of Thompson et al. [18], which was modified to some extent to accommodate various Gaussian FWHM parameters and the Scherrer co-efficient P for Gaussian broadening. The asymmetric parameters are refined using the Berar-Baldinozzi function employing the multi-beam Simpson rule integration devised by Howard [19]. A correction for preferred orientation of the crystallites in the sample is dealt with the model as proposed by March-Dollase [20, 21]. The Legendre polynomial of first kind was used to fit the background. The fitted profile of the observed and calculated relative X-ray intensities along with their differences for calcite is shown in figure 3. The atomic positional parameters, displacement parameters, cell parameters, thermal vibration parameters, refined bond distance and structural parameters from Rietveld method with the corresponding standard deviations in parenthesis are given in tables 1 and 2.

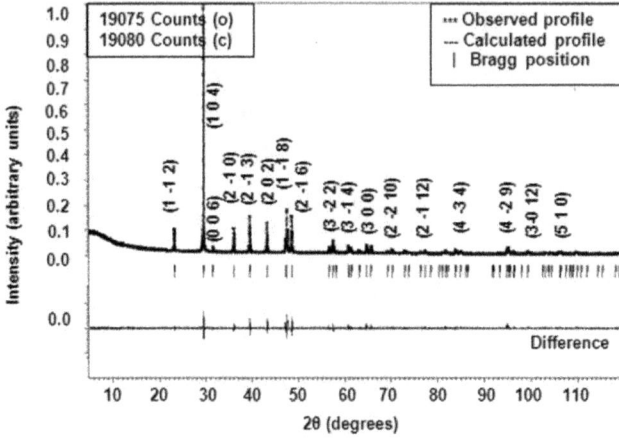

Figure 3. The fitted powder XRPD profile of calcite.

Table 1. Atomic positions and thermal vibration parameters of calcite.

Parameter	Ca	C	O
x	0	0	0.2534(2)
y	0	0	0
z	0	0.25(1)	0.25(2)
U_{11}	0.0164(8)	0.0138(5)	0.0132(5)
U_{22}	0.0164(8)	0.0138(5)	0.0276(6)
U_{33}	0.0172(5)	0.0129(6)	0.0274(5)
U_{12}	0.0086(4)	0.0033(4)	0.0146(2)
U_{13}	0	0	- 0.0051(3)
U_{23}	0	0	-0.0062(5)

Table 2. Refined bond distances and structural parameters of calcite.

Parameter	Value
a (Å)	4.9912 (9)
c (Å)	17.0717(11)
Cell Volume(Å^3)	368.1373(44)
Density (gm/cc)	2.7079(3)
R_p (%)	7.47
R_{obs} (%)	3.56
GoF	1.34

4. CHARGE DENSITY ANALYSIS

The maximum entropy method (MEM) [14] is a great tool to evaluate accurate charge density. The observed structure factors of calcite extracted from the Rietveld [13] method were used for the MEM procedure to obtain the charge density distribution in the unit cell. To evaluate the charge density, the software PRIMA [22] was used, which employs the maximum entropy method proposed by Collins [14].

The resultant electron density is plotted with the help of the visualization software VESTA [23]. In the present calculation, the unit cell was divided into 128×128×128 pixels along the unit cell. The uniform prior density was used by dividing the total number of electrons by the volume of the unit cell. The obtained MEM [14] parameters of calcite have been given in table 3. In this table, R_{MEM} represents the reliability index from the MEM [14] refinements and wR_{MEM} represents the weighted reliability index.

Figure 4(a) represents the individual atoms imposed on the unit cell of calcite and the iso-surface levels are suppressed for a better view. Three dimensional electron density distribution on the unit cell of calcite is shown in figure 4(b) considering the iso-surface level of 0.15 e/Å^3. Different view (half the unit cell along z-axis) of calcite with the same iso-surface level is shown in figure 4(c) in order to visualize the bonding nature between the atoms. Figure 5(a) shows all the trigonal planar CO_3 groups consisting of carbon atom surrounded by three oxygen atoms. Another view (one-fourth of the unit cell along the z-axis) with the trigonal planar CO_3 groups of atom in calcite structure is shown in figure 5(b). The trigonal planar CO_3 groups are constrained to be parallel and constitute a plane perpendicular to the c-axis in the unit cell of calcite and the trigonal planar CO, groups thus formed within the unit cell of calcite are more symmetrical. This is a typical feature of carbonate crystal structure, determining strong anisotropy of their physical properties.

Each Ca atom is coordinated by six O atoms at the vertices of an octahedron. The bond length between C – O in the planar CO_3 and Ca – O in octahedral structure of calcite is found to be of 1.2647 Å and 2.3695 Å respectively. The Ca atom is in octahedral coordination with -3 symmetry (one-eighth of the unit cell of calcite along the z-axis) is shown in figure 5(c). Each O atom is shared between two octahedra and also forms one corner of an equilateral triangle, perpendicular to the c-axis, with the carbon atom at its center in the unit cell of calcite as shown in figure 5(d). This causes the groups to align themselves between the layers of cations in such a way that each O in the group is attached to a cation from above and below with equal bonding distance that balance each other. Any applied electric field that causes the rotation and the vertical motion of the group breaks the high symmetry configuration with an increased energy of the crystal, thus resulting in non-switching (non-ferroelectric) behaviour which plays a significant role in the birefringence property of the calcite structure.

3D outlook of the unit cell of calcite with the (001) plane is shown in figure 6(a) and figure 6(b) shows the different view of the same (001) plane which is passing through the Ca atoms in the perpendicular direction. Figure 6(c) shows the 2D electron density distribution corresponding to the (001) plane. The contour range is from 0.0 e/Å^3 to 3.0 e/Å^3 and the step size is 0.12 e/Å^3. 3D view of the unit cell of calcite with the (001) plane at a distance of 1.505 Å away from the origin of the cell is shown in figure 7(a). The two-dimensional electron density distribution on this plane is shown in figure 7(b). The contour range is from 0.0 e/Å^3 to 3.0 e/Å^3 and the step size is 0.17 e/Å^3. The covalent bonding nature between oxygen and carbon within the trigonal planar CO_3 group has been visualized through the contour lines in both the figures 7(a) and 7(b).

Figure 8(a) shows the 3D view of the unit cell of calcite with the (010) plane and the 2D electron density distribution on that plane with the contour range of 0.0 e/Å^3 1.8 e/Å^3 with the step size 0.18 e/Å^3 is shown in figure 8(b). 3D view of the unit cell of calcite with the (101) is shown in figure 9(a). 2D electron density distribution on that plane with the contour range of 0.0 e/Å^3 to 2.0 e/Å^3 and contour interval of 0.3 e/Å^3 is shown in figure 9(b). Figure 10(a) shows the unit cell of calcite with the (110) plane. 2D electron density distribution on that plane with the contour range of 0.0 e/Å^3 to 1.8 e/Å^3 and contour interval 0.24 e/Å^3 is shown in figure 10(b). The qualitative understanding and the analysis can be done by visualizing the one dimensional profiles of the charge density along the bonding directions. Hence one-dimensional profiles of electron density are drawn along Ca–O and C–O directions and are shown in figures 11(a) and 11(b) respectively. The ionic and covalent nature of bonding was revealed in Ca–O and C–O respectively. The first atom has been considered to be in the origin in the above mentioned one-dimensional electron density profiles.

In order to understand the charge density distribution and the bonding features the multipole model proposed by Hansen et al. [24] was used. This model describes the atomic charge density as a series expansion in real spherical harmonic functions through the fourth order Y_{lm}. This model also describes the charge density in a crystal as the superposition of harmonically vibrating aspherical atomic density distributions convolved with the Gaussian thermal displacement distribution term. To analyze the charge density in detail the topological view of the charge distribution is done using the atoms in the molecules theory as proposed by Bader [25]. According to him, two atoms are bonded if they are connected by a line of maximum electron density called a bond path, on which lies a bond critical point (BCP) where $\vec{\nabla}\rho(r_{BCP})=0$, and the critical points are the characteristics of the bonding existing between the atoms. The multipole model refinement for the chosen material is done using the software JANA 2006 [17] which accommodates neutral atom wave function from Clementi [26] tables and Slater type radial functions for the construction of core and valence charge density of the atoms Ca, C and O. The constructed charge densities were analyzed and the critical points are searched using the Newton - Raphson method. The refined multipole parameters are given in table 4. The (r_{BCP}) using bond critical points (BCP) defined as (3, -1) are listed in table 5. The obtained values were compared with the limit defined by Tsierelson [27]. According to him the closed shell interactions will have a low (r_{BCP}), i.e., 0.07 e/Å^3 < (r_{BCP}) < 0.25 e/Å^3, and $0.12 < |\lambda_1|/\lambda_3 < 0.17$, and for shared shell interactions (r_{BCP}) will be high, i.e., 0.30 e/Å^3 < (r_{BCP}) < 2.0 e/Å^3, $|\lambda_1|/\lambda_3 > 0.17$, and $\nabla^2(r_{BCP}) < 0$.

In our work, we have compared the information on bond critical points (BCP) derived using both the multipole method and maximum entropy method (MEM) [14]. The numerical values of the mid band electron densities between different atoms of calcite from one-dimensional MEM analysis are given in table 5. The values obtained are very close to each other and there seem to be no major deviations in the models adopted. The difference in (r_{BCP}) between bond critical points (BCP) and the maximum entropy method (MEM) [14] model is due to the inherent inability for the multipole model which depends on many parameters and also on the accuracy with which the Fourier coefficients are being determined. However the maximum entropy method (MEM) [14] model is unbiased and accurate due to its dependence on the accurate input information. Unlike the multipole model, the maximum entropy method (MEM) [14] does not suffer from termination error and hence it is capable of producing an all positive charge density which can be relied on for the understanding of any bonding features. As found in table 5 (r_{BCP}) between C–O bond enacts a perfect covalent bond. This is mainly due to the difference in electronegativity O and C (0.89) resulting in the contribution of covalent character to the bond as 81%. The (r_{BCP}) between Ca–O shows the bond is ionic with

21% of covalent character mixed with it. This results in (r_{BCP}) being close to the lower end limit of covalent character as defined by Tsierelson [27]. All these analyses show the reliability of the maximum entropy method (MEM) [14] model over any other methodology usually adopted for the charge density distribution.

a)   b)   c)

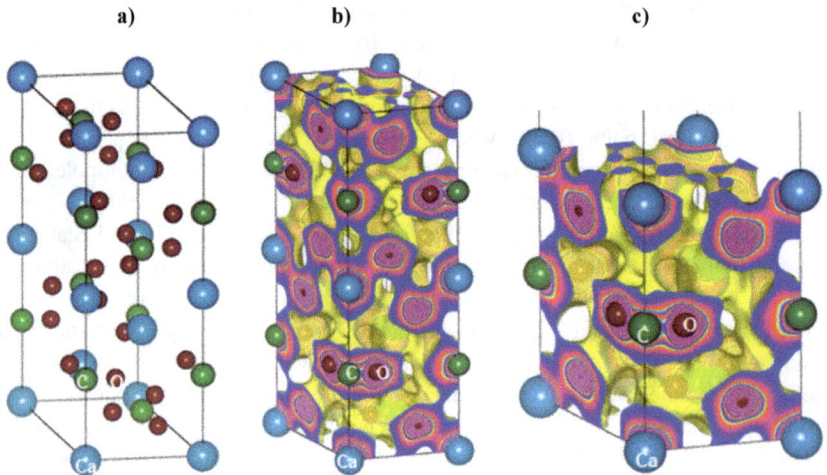

Figure 4(c). 3D view (half the unit cell along z-axis) of electron density of calcite with the iso-surface level 0.15 e/$Å^3$.

a)

b)

c)

d)

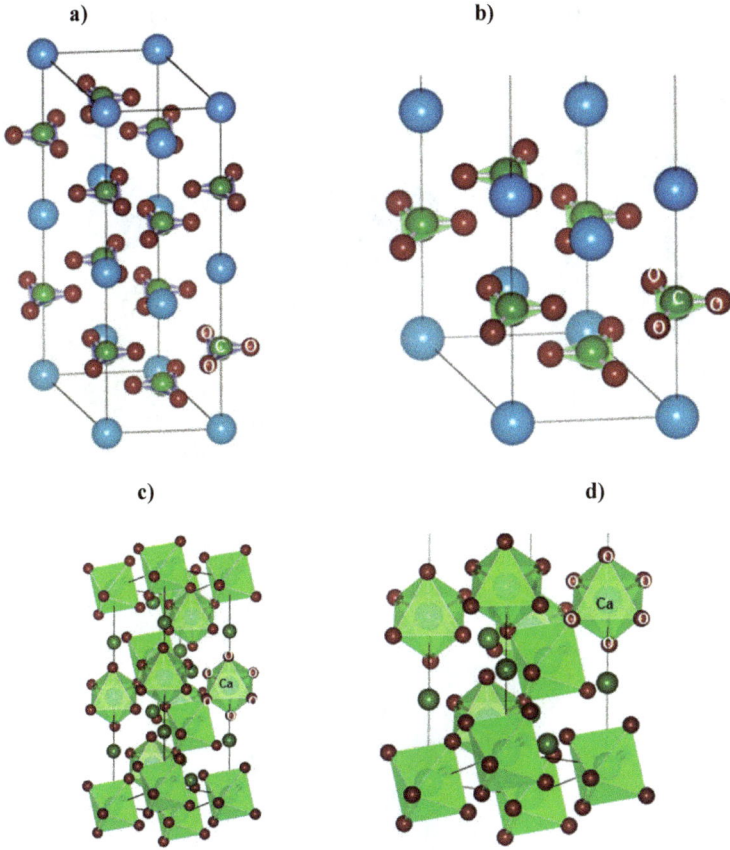

Figure 5(a). Unit cell of calcite containing trigonal planar CO_3. (b). 3D representation (one-fourth of the unit cell along z-axis) of trigonal planar CO_3 groups in the unit cell of calcite. c) 3D view octahedral structure (one-eighth of the unit cell along z-axis) of Ca with O atoms in the unit cell of calcite. d) Unit cell of calcite containing CaO_6 octahedra and trigonal planar CO_3.

a)

b)

c)

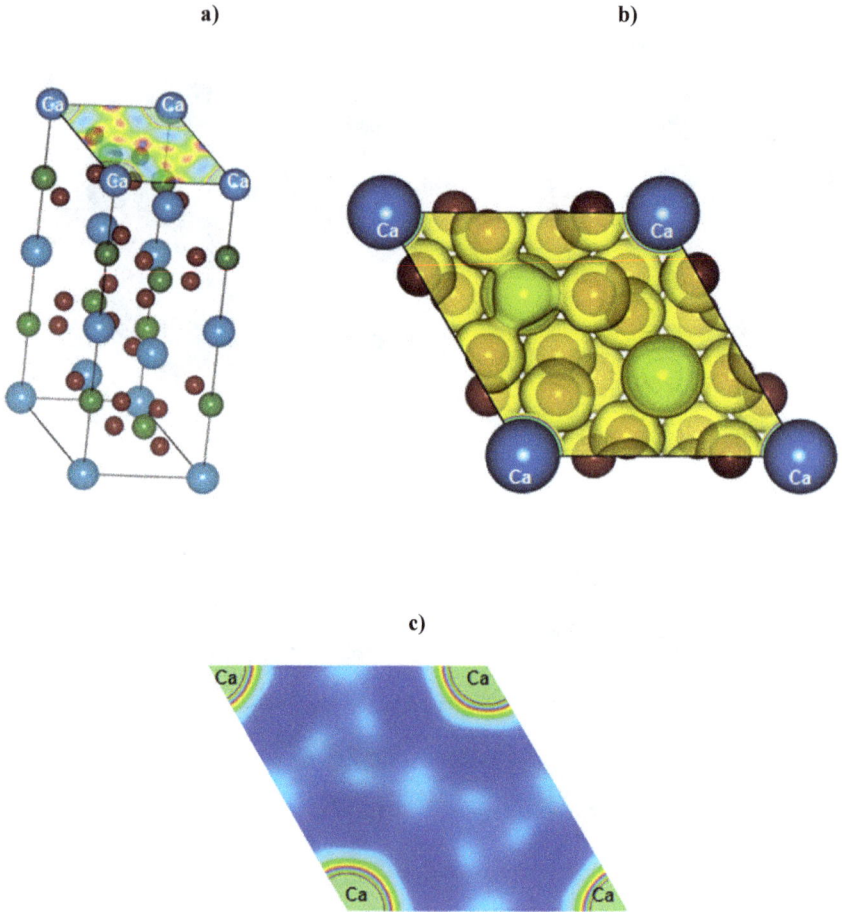

Figure 6(a). The (001) plane passing through Ca atoms in calcite. (b). 3D electron density of calcite with (001) plane passing through Ca atoms in the perpendicular direction. (c). 2D MEM electron density distribution of calcite on (001) plane passing through Ca atoms of Calcite (Contour range is from 0 e/$\mathring{A}^3$ to 3.0 e/$\mathring{A}^3$ and Contour interval is 0.12 e/$\mathring{A}^3$).

a)

b)

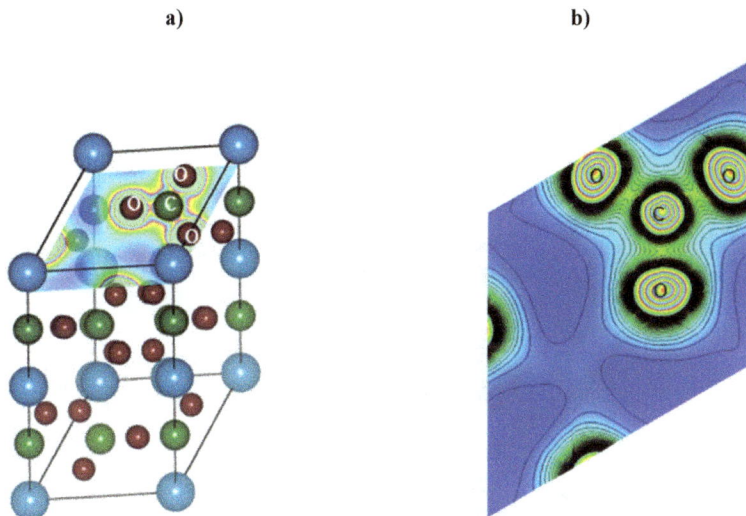

Figure 7(a). The (001) plane passing through C and O atoms in calcite at a distance of 1.505 Å from the origin in the unit cell of calcite. (b). 2D MEM electron density distribution on (001) plane of calcite at a distance of 1.505 Å from the origin (Contour range is from 0 e/Å³ to 3.0 e/Å³ and Contour interval is 0.17 e/Å³)

a)

b)

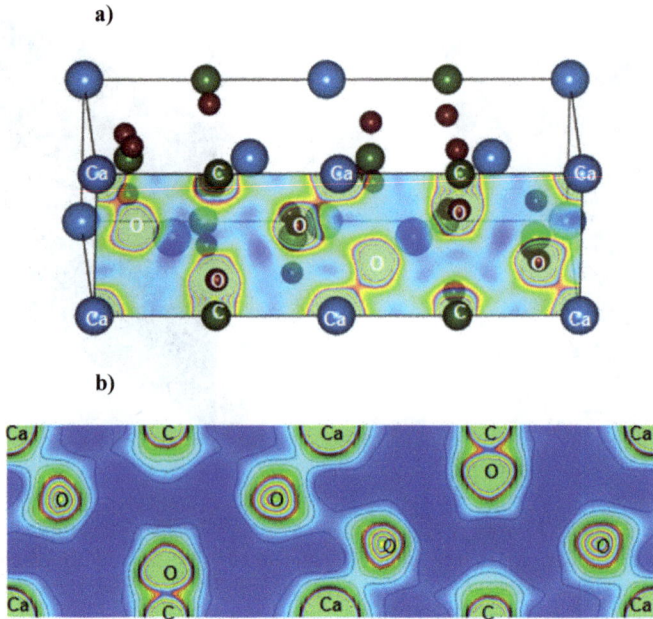

Figure 8(a). The (010) plane passing through Ca, C and O atoms in the unit cell of calcite. (b). 2D MEM electron density distribution on (010) plane of calcite (Contour range is from 0 e/Å³ to 1.8 e/Å³ and Contour interval is 0.18 e/Å³).

a)

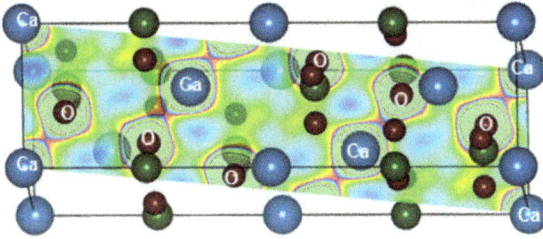

b)

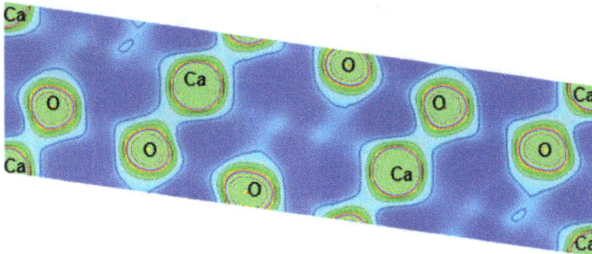

Figure 9(a). The (101) plane passing through Ca and O atoms in the unit cell of calcite. (b). 2D MEM electron density distribution on (101) plane of calcite (Contour range is from 0 e/$Å^3$ to 2.0 e/$Å^3$ and Contour interval is 0.3 e/$Å^3$).

)

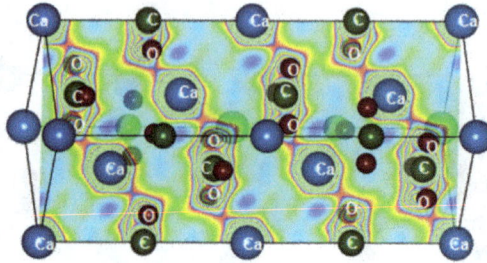

b)

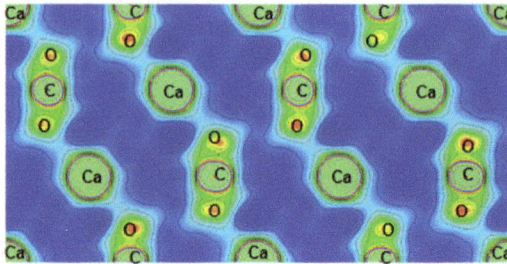

Figure 10(a). The (110) plane passing through Ca, C and O atoms in the unit cell of calcite. (b). 2D MEM electron density distribution on (110) plane of calcite (Contour range is from 0 e/Å³ to 1.8 e/Å³ and Contour interval is 0.24 e/Å³).

a)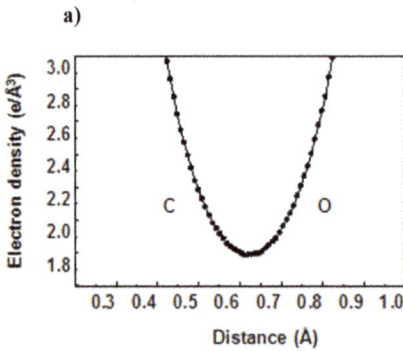

b)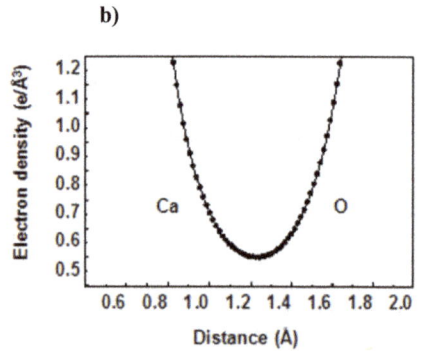

Figure 11(a). One-dimensional variation of electron density along Ca – O bond. (b). One-dimensional variation of electron density along C – O bond.

Table 3. Parameters from MEM refinements of calcite.

Parameter	Value
Number of cycles	3051
Lagrange parameter (λ)	0.0040
Number of electrons/unit cell(F_{000})	300
R_{MEM} (%)	1.1443
wR_{MEM} (%)	1.8522

Table 4. Parameters from Multipole refinement.

Parameter	Ca	C	O
P_c	17.97736	2.016297	2.000387
P_v	2.233656	4.075812	6.167906
κ	1.136618	0.971690	0.988666
P_{20}	0.154843	0.149570	0.103540

Table 5. Analysis on (3, -1) bond critical point and MEM bond density.

Parameter	Multipole	MEM
Ca – O		
Ca to (3,-1) (Å)	1.2605(5)	1.2245
(3,-1) to O (Å)	1.1085(4)	1.1450
r(Ca- O) (Å)	2.3690(7)	2.3695
$\rho(r_{BCP})$ (e/Å^3)	0.4063	0.4994
C – O		
C to (3,-1)(Å)	0.6363(8)	0.6112
(3,-1) to O (Å)	0.6324(7)	0.6535
r(C- O) (Å)	1.2687(9)	1.2647
$\rho(r_{BCP})$ (e/Å^3)	1.8992	1.7856

5. CONCLUSION

The charge density analysis for calcite was done using the MEM technique and it was compared with the multipole analysis. The charge density distribution in calcite shows the predominant ionic along with covalent bond existing between C – O. The super resolution density maps plotted with the adopted MEM technique give qualitative as well as quantitative picture of the electronic structure of the chosen material. The morphological and optical analyses were also done using scanning electron microscopy and UV-Visible spectrophotometer.

ACKNOWLEDGEMENT

The authors are grateful to The Madura College (Autonomous), Madurai, Tamilnadu, India and SAIF, Cochin University, India for extending their experimental facilities.

REFERENCES

[1]   E. M. Landau, M. Levanon, L. Leiserowitz, M. Lahav and J. Sagiv, Transfer of structural information from Langmuir monolayers to three-dimensional growing crystals, Nature 318 (1985) 353-356.
http://dx.doi.org/10.1038/318353a0

[2]   S. Mann, B. R. Heywood, S. Rajam and J.D. Birchal, Controlled crystallization of $CaCO_3$ under stearic acid monolayers, Nature 334 (1988) 692-695.
http://dx.doi.org/10.1038/334692a0

[3]   J. Aizenberg, A. J. Black and G. M. Whitesides, Controlling local disorder in self-assembled monolayers by patterning the topography of their metallic supports, Nature 394 (1998) 868-871.
http://dx.doi.org/10.1038/29730

[4]   G. Falini, S. Albeck, S. Weiner and L. Adadi, Control of Aragonite or Calcite Polymorphism by Mollusk Shell Macromolecules, Science 271 (1996) 67-69.
http://dx.doi.org/10.1126/science.271.5245.67

[5]   D. Chakrabarty and S. Mahapatra, Aragonite crystals with unconventional morphologies, J. Mater. Chem. 9 (1999) 2953-2957.
http://dx.doi.org/10.1039/a905407c

[6]   K. Naka, D.K. Keum, Y. Tanaka and Y. Chujo, Control of crystal polymorphs by a 'latent inductor': crystallization of calcium carbonate in conjunction with *in situ* radical polymerization of sodium acrylate in aqueous solution, Chem. Commun. 16 (2000) 1537-1538.

http://dx.doi.org/10.1039/b004649n

[7]   N.A.J. M. Sommerdijk and G. With, Biomimetic $CaCO_3$ Mineralization using Designer Molecules and Interfaces, Chem. Rev. 108 (2008) 4499-4550.
http://dx.doi.org/10.1021/cr078259o

[8]   E.N. Maslen, V.A. Streltsov and N.R. Streltsova, X-ray study of the electron density in calcite, $CaCO_3$, Acta Cryst. B49 (1993) 636-641.
http://dx.doi.org/10.1107/S0108768193002575

[9]   G. Wullf, On the question of the speed of growth and dissolution of Krystallflächen, Z. Krystallogr. 34 (1901) 449-530.

[10]   P.W. Bridgman, The high pressure behavior of miscellaneous minerals, Am. J. Sci. 237 (1939) 7-18.
http://dx.doi.org/10.2475/ajs.237.1.7

[11]   A.K. Singh and G.C.Kennedy, Compression of calcite to 40 KB. J. Geophys. Res. 79 (1974) 2615-2622.
http://dx.doi.org/10.1029/JB079i017p02615

[12]   L. Merrill and W.A. Bassett, The crystal structure of $CaCO_3$ (II), a high-pressure metastable phase of calcium carbonate, Acta Cryst. B31 (1975) 343-349.
http://dx.doi.org/10.1107/S0567740875002774

[13]   H. M. Rietveld, A profile refinement method for nuclear and magnetic structures, J. Appl. Crystallogr. 2 (1969) 65-71.
http://dx.doi.org/10.1107/S0021889869006558

[14]   D.M. Collins, Electron density images from imperfect data by iterative entropy maximization, Nature 298 (1982) 49-51.
http://dx.doi.org/10.1038/298049a0

[15]   D.L. Wood and J. Tauc, Weak Absorption Tails in Amorphous Semiconductors, Phys. Rev. B5 (1972) 3144-3151.
http://dx.doi.org/10.1103/PhysRevB.5.3144

[16]   M.K. Aydinol, J.V. Mantese and S.P. Alpay, A comparative *ab initio* study of the ferroelectric behaviour in KNO_3 and $CaCO_3$, J. Phys. Condens. Matter. 19 (2007) 496210.
http://dx.doi.org/10.1088/0953-8984/19/49/496210

[17]   V. Petříček, M. Dušsek and L.Palatinus, Crystallographic Computing System JANA2006: General features. Z. Kristallogr. 229(5) (2014) 345-352.
http://dx.doi.org/10.1515/zkri-2014-1737

[18]   P. Thompson, D.E. Cox and J.B. Hastings, Rietveld refinement of Debye-Scherrer synchrotron X-ray data from Al_2O_3, J. Appl. Cryst. 20 (1987) 79-83. http://dx.doi.org/10.1107/S0021889887087090

[19]   C.J. Howard, The approximation of asymmetric neutron powder diffraction peaks by sums of Gaussians, J. Appl. Crystallogr. 15 (1982) 615-620. http://dx.doi.org/10.1107/S0021889882012783

[20]   A. March, Mathematische Theorie der Regelung nach der Korngestah bei affiner Deformation, Z. Kristallogr. 81 (1932) 285-297. http://dx.doi.org/10.1524/zkri.1932.81.1.285

[21]   W.A.J. Dollase, Correction of intensities for preferred orientation in powder diffractometry: application of the March model, J. Appl. Cryst. 19 (1986) 267-272. http://dx.doi.org/10.1107/S0021889886089458

[22]   F. Izumi and R.A. Dilanian, Recent research developments in physics. Transworld research network Vol. 3, Part II, Trivandrum, (2002) 699.

[23]   K. Momma and F. Izumi, *VESTA*: a three-dimensional visualization system for electronic and structural analysis, J. Appl. Cryst. 41 (2008) 653-658. http://dx.doi.org/10.1107/S0021889808012016

[24]   N.K. Hansen and P. Coppens, Testing aspherical atom refinements on small-molecule data sets. Acta Cryst. A34 (1978) 909-921. http://dx.doi.org/10.1107/S0567739478001886

[25]   R.F.W. Bader, 1990. Atoms in Molecules-A Quantum Theory. Oxford University Press, Oxford.

[26]   E. Clement and C. Roetti, Roothaan-Hartree-Fock atomic wavefunctions: Basis functions and their coefficients for ground and certain excited states of neutral and ionized atoms, $Z \leq 54$, Atomic data and nuclear tables 14 (1974) 177-478.

[27]   V.G. Tsierelson, Acta Cryst. A55 supplement, Abstract M13-OF-003, 1999.

CHAPTER 11

Synthesis and Characterization of $NiFe_2O_4$ Nano Particles Prepared by the Chemical Reaction Method

Y.B.Kannan[a], R.Saravanan[b], N.Srinivasan[c].

[a]Department of Physics, Arumugam Pillai Seethai Ammal College, Tiruppattur – 630 211, India.

[b]Research Centre & PG Department of Physics, The Madura College, Madurai – 625 011, India.

[c] Research Centre & PG Department of Physics, Thiagarajar College, Madurai – 625 009, India.

Email: ybkans@gmail.com; saragow@gmail.com; vasan692000@yahoo.co.in

Abstract

The present study was carried out to determine the numerical values of the bond strength using the maximum entropy method (MEM) between atoms at various sites, namely tetrahedral – tetrahedral, tetrahedral – octahedral and octahedral – octahedral sites, of interactions in $NiFe_2O_4$ nanoparticles prepared by solid state reaction. The experimental lattice parameter agrees well with theoretical lattice parameters. The particle size lies in the nanometer regime. SEM reveals the presence of porosity in the sample. The EDAX and the XRF analysis confirm the elemental composition and purity of the samples. Interestingly, due to the presence of the Yaffet-Kittel angle, instead of the octahedral - octahedral site interactions, the tetrahedral – tetrahedral site interactions remain the strongest interaction in this study. The hysteresis curve together with small coercivity value reveals the presence of small magnetic particles exhibiting a super paramagnetic behavior. The dielectric constant determined through broad band dielectric spectrometer (BDS) shows a normal behavior whereas the dielectric loss tangent shows abnormal behavior. The band gap energy of 2.1eV is evaluated from the optical study.

Keywords

Solid State Reaction Method; Spinel Ferrites; X-Ray Diffraction; Rietveld Analysis; Electron Charge Density using MEM Studies

Contents

1.   INTRODUCTION

Nanostructured spinel type ferrites, having the general formula of MFe_2O_4 where M is any divalent ion of metals such as nickel, cadmium, zinc, magnesium, copper, etc., are promising materials which could be used as susceptors of induction heating in catalytic chemical reactors due to their high Curie temperatures and moderate magnetic losses in the kHz range. Moreover, they can be employed as magnetic catalysts or catalyst carriers which facilitate separation. [1]. Ferrite nanocrystals are also of interest in various applications, such as inter-body drug delivery, bioseparation, and magnetic refrigeration systems [2]. Structural, electrical, and magnetic properties of these materials effectively depend upon their stoichiometry, methods of synthesis, and cationic distributions among the available tetrahedral A and octahedral B sites of the face-centered cubic (*fcc*) spinel structure formed by oxygen anions at the corners. [3]. The change in cation distribution may result in unexpected electrical and magnetic behaviour [4]. Nickel and substituted nickel ferrite are versatile and technologically important soft ferrite materials because of their typical ferromagnetic properties, low conductivity and thus lower eddy current losses, high electrochemical stability, catalytic behaviour, abundance in nature, etc. [5].

As far as the authors are concerned, there is a lack of investigations on ferrite materials regarding the bond strength between the atoms at tetrahedral A site and octahedral B site by employing the maximum entropy method (MEM), from which numerical values for the various site interactions, namely A-A, A-B and B-B, could be derived. Hitherto, only theoretical predictions of these site interactions were published. Hence, in this work, we evaluated mid-bond electron density values for various site interactions using the maximum entropy method (MEM) and report the same along with their structural, optical, dielectric and magnetic studies of $NiFe_2O_4$ nano ferrite particles.

2. SAMPLE PREPARATION

Nickel Ferrite nano particles were prepared using the chemical reaction method. Analytical grade nickel sulphate ($NiSO_4.6H_2O$), iron nitrate ($Fe(NO_3)_3.9H_2O$), sodium hydroxide (NaOH) and sodium chloride (NaCl) all from Alpha Aeser Chemicals, were used to prepare the sample. The molar ratio of $NiSO_4.6H_2O$, $Fe(NO_3)_3.9H_2O$, NaOH and NaCl is 1:2:8:10 and ground together in an agate mortar pestle for about 1 h. The reaction takes place exothermally and the color changes from greenish red to brown during this mixing process. This mixture was subjected to calcinations at 700 °C for 1 h. The powder was crushed and then washed with deionized water to remove sodium chloride and dried at 100 °C for 1 h to obtain polycrystalline nickel ferrite nano particles. The samples were pressed into disc shaped pellets with a thickness of 0.13 cm and radius of 1.3 cm by applying pressure for dielectric measurements.

X-ray diffraction pattern of the sample was recorded at room temperature using a Bruker AXS D8 advance X-ray diffractometer with Cu radiation ($\lambda = 1.5406$ Å). The sample was exposed to the radiation with a primary beam power of 40 kV and 35 mA with step scan of 0.02° in the 2θ range of 20°-120°. Optical absorption spectra of the samples were recorded in the UV-vis wavelength range of 2000 - 7500 Å. The magnetic properties of the samples were studied at room temperature by using a vibrating sample magnetometer (Lakeshore VSM 7410 Model). Dielectric properties of the sample were studied at room temperature in the frequency range of 0.1Hz to 15MHz using NOVOCONTROL Technologies GmbH & Co, Germany, (Model: Concept 80).

3. RESULTS AND DISCUSSION

3.1 X-RAY DIFFRACTION ANALYSIS

The XRD data of $NiFe_2O_4$ were collected at room temperature and the peaks in the raw X-ray diffraction pattern (not shown) have been indexed with spinel structure with Fd-3m (227) space group (JCPDS. file no. 10-0325). The broad XRD peaks indicate that the particles are in the nano meter regime. The low intensity peaks reveal poor crystalline nature of the sample. No other reflection peaks related to a secondary phase are found in the XRD confirming the formation of a single phase. The cation distribution in the present study is obtained from the analysis of the XRD data. The calculated intensity of the planes (I_{hkl}) is calculated using the following relation [6]

$$I_{hkl} = |F_{hkl}|^2 \, P \, L_p \tag{1}$$

where F is the structure factor, P is the multiplicity factor and L_p is the Lorentz polarization factor. It is well known that the intensities of plane (220) and (440) are sensitive to cation on the A site, whereas the (400) and the (422) plane intensities are sensitive to cation on the B site [6] hence, the X-ray intensity ratios of I_{220}/I_{400}, I_{220}/I_{440}, I_{422}/I_{440} and I_{400}/I_{440} were used for various cation distribution combinations and those matching closely with the observed intensity ratios and corresponding cation distributions are listed in Table 1.

Table 1. Cation distribution and X-ray intensity ratio of $NiFe_2O_4$

Cation distribution	I_{220}/I_{440}		I_{400}/I_{440}		I_{220}/I_{400}	
	Obs	Cal	Obs	Cal	Obs	Cal
$(Ni_{0.065}Fe_{0.935})^A[Ni_{0.935}Fe_{1.065}]^BO_4$	0.20	0.21	0.76	0.80	0.27	0.27

Using this cation distribution as occupancy factor over the available A and B sites in the unit cell, the Rietveld method [7] is employed for refining the background, pseudo-voigt, asymmetry, preferred orientation, scale and lattice parameters using the software Jana 2006 [8]. The refined powder profile of the sample is shown in Fig.1 and all the structural parameters along with magnetic parameters are listed in Table 2.

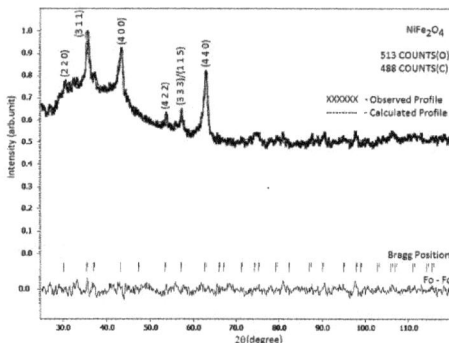

Figure 1. Rietveld refined powder profile of $NiFe_2O_4$.

Table 2. Structural and MagneticParameters of the Sample NiFe₂O₄

Parameters			Value
Lattice parameter	Theoretical	a_{th}(Å)	8.3116
	Experimental	a_{exp}(Å)	8.3512(2)
Cell volume		(Å)³	582.4
X-ray density		D_x(g/cm³)	5.346
Bulk density		D_m(g/cm³)	3.210
Average crystallite size		t(nm)	9
Oxygen positional parameter		'u'	0.3791(4)
Radius of tetrahedral	From cation distribution		0.6432
site r_A(Å)	From XRD		0.5480
Radius of octahedral	From cation distribution		0.6638
site r_B(Å)	From XRD		0.7332
Tetrahedral shared edge length		d_{AES}(Å)	3.0504
Octahedral shared edge length		d_{BES}(Å)	2.8547
Octahedral unshared edge length		d_{BEU}(Å)	2.9534
Saturation Magnetization (emu/g)		M_S	11.6
Coercivity (H_{Ci}) G			3.9
Retentivity (emu/g)		(M_r)	48.28
Bohr	μ_B^H (From hysteresis loop)		0.48
Magneton	μ_B^N (From cation distribution)		2.39
Yaffet-Kittel angle (°)			42

The lattice parameter (a_{exp}) value is found to be 8.35 Å. Various other authors [2, 3] also reported the same range of value. The X-ray density ((D_x), measured bulk density (D_m), porosity (P) and the average particle size (t)) are calculated using the formula [3]. The D_x value is higher than that of the D_m value; this indicates the presence of pores in the sample. The full width half maximum (FWHM) values of the (311), (400) and (440) peaks were considered in calculating the average particle size of the sample which is found to be 9 nm.

The theoretical lattice parameter (a_{th}) value is theoretically calculated using the equation [3]

$$a_{th} = \frac{8}{3\sqrt{3}}(r_A + R_o) + \sqrt{3}\,(r_B + R_o) \tag{2}$$

where R_o (1.32 Å) is the radius of the oxygen ion, r_A and r_B are the radius of the A and B site ions. The value of r_A and r_B can be calculated from cation distribution (Table 1), with the following equations [3]

$$r_A = (0.065r_{Ni}^{2+} + 0.935r_{Fe}^{3+})$$ (3)

$$\text{and } r_B = \frac{1}{2}[0.935r_{Ni}^{2+} + 1.065\ r_{Fe}^{3+}]$$ (4)

For nickel and iron ions, the ionic radius values are taken as $r_{Ni}^{2+} = 0.69$ Å and $r_{Fe}^{3+} = 0.64$ Å. The a_{th} value is calculated as 8.31 Å and its close agreement with a_{exp} validates the cation distribution in the present study. r_A and r_B values can also be evaluated from XRD data using equations [3]

$$r_A = a_{exp} (\sqrt{3})(u - 0.25) - R_o.$$ (5)

$$r_B = a_{exp} (5/8 - u) - R_o.$$ (6)

Small difference exists between the r_A and r_B value, obtained from XRD and cation distribution as shown in Table 2, which may be attributed to the presence of Fe^{2+} ions ($r_{Fe}^{2+} = 0.78$ Å) on the B sites and these Fe^{2+} ions force the Fe^{3+} ions to migrate to the A sites. Since the bigger Fe^{2+} ions ($r_{Fe}^{2+} = 0.78$ Å) occupy the B sites, the radius of the B site (r_B) is greater than that of (r_A) which is attributed to the small difference that exists between a_{th} and a_{exp} [3].

3.2 SEM, EDAX AND XRF ANALYSIS

Fig. 2a shows the scanning electron microscope image of the sample. Random distribution in the particle sizes and presence of pores can be clearly seen. The EDAX analysis of the sample is shown in Fig. 2b. The expected stoichiometry for nickel to iron is 0.5 whereas the estimated stoichiometry of that from EDAX analysis comes out as 0.4 only. The elemental composition of the ferrite sample was analyzed by the X-ray fluorescence (XRF) method and the % concentrations of Ni, Fe and O were found to be 31.68, 34.72 and 26.5 respectively. The purity of the sample is confirmed by the presence of nickel, iron and oxygen and not any other intermediate reactions product.

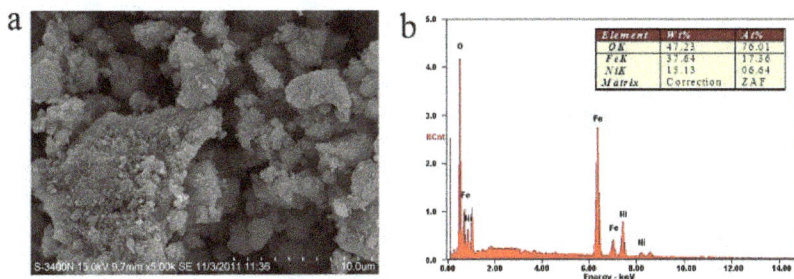

Element	W't%	At%
O K	47.23	76.01
Fe K	37.64	17.36
Ni K	13.13	06.64
Matrix	Correction	ZAF

Figure 2. a). SEM micrograph b) EDX spectrum of NiFe$_2$O$_4$.

3.3 ELECTRONIC CHARGE DISTRIBUTION STUDIES USING MEM

MEM was introduced by Gull and Daniel [9] and Collins [10] who formulated this method for crystallographic applications. The structure factors extracted from the Rietveld refinement method [7] are used in the maximum entropy method (MEM). MEM is an exact tool to study the electron density distribution because of its resolution. The bonding nature and the distribution of electrons in the bonding region can be clearly visualized using this technique [9]. For the numerical MEM computations, the software package PRIMA [11,12] is used. For the 3D, 2D and 1D representation of the electron densities, the program VESTA package was used [13]. The MEM refinements were carried out by dividing the unit cell into $64 \times 64 \times 64$ pixels. The initial electron density at each pixel is fixed uniformly as $F_{000}/a_{exp}^3 = 1.538$ e/Å^3, where F_{000} is the total number of electrons in the unit cell and a_{exp} is the cell parameter. The three dimensional MEM electron density distribution of the sample under investigation is shown in Fig. 3 in which 'A' represents the tetrahedral sites, 'B' represents the octahedral sites and 'O' represents the oxygen atoms. The iso-surface level is suppressed for a better view in Fig. 3a.

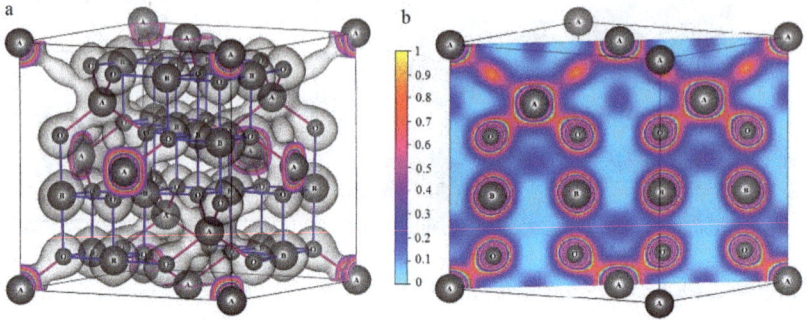

Figure 3. a) 3D electron density of $NiFe_2O_4$. b) 2D electron density of $NiFe_2O_4$ on (110) plane. Contours lines are drawn between 0 to $1e/Å^3$ with $0.1e/Å^3$ interval.

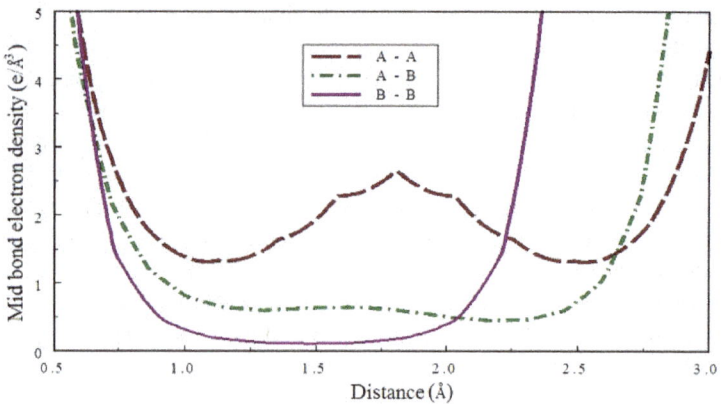

Figure 4. 1D electron density between $A - A$, A-B and B-B site atoms in the unit cell of $NiFe_2O_4$.

2D electron density distribution on the miller plane (110) of the sample in the density range of $0 - 1.0$ $e/Å^3$ with contour interval of 0.1 $e/Å^3$ is drawn using the software program VESTA [13] and shown in Fig. 3b. In Ferrites, the A-B interaction takes place through super exchange interaction i.e., the atoms at the A sites and at the B sites interact through oxygen atoms. The super exchange interaction between atoms at the A-site and at B-site through oxygen atoms in (110) plane is clearly seen and also the interactions between the two nearest A-sites atoms in the (110) plane can also be seen by the presence of the residual electron cloud in Fig 3b. The mid bond electron density values give the strength of the A-B, A-A and B -B interactions of the sample, hence 1D the electron density view is drawn and the same is shown in Fig. 4 and the values are tabulated in Table 3. It can be seen that the mid bond electron density value of A-A interaction is the strongest, followed by A-B and B-B interaction values. A-B interactions not being the strongest suggests that non-collinear spin exists in the unit cell. Interestingly, the value of A-A interactions is more than that of the B-B interactions and it may be attributed to the presence of Ni^{2+} ions at the A site.

Table 3 Mid-bond electron density values between A and B sites of $NiFe_2O_4$

A – A bond		A – B bond		B – B bond	
Mid bond electron density	Distance	Mid bond electron density	Distance	Mid bond electron density	Distance
2.64 e/Å³	1.80 Å	0.64 e/Å³	1.63 Å	0.11 e/Å³	1.47 Å

Table 4 Interatomic distances between cations (M_e – M_e), between anions – cations (M_e – O) and interatomic angles (θ) of the sample

M_e – M_e (Å)					M_e – O (Å)				θ (°)				
b	c	d	e	f	p	q	r	s	θ_1	θ_2	θ_3	θ_4	θ_5
2.95	3.46	3.62	5.42	5.11	2.05	1.87	3.58	3.64	123.9	147.7	91.9	125.7	76.1

The configuration of ion pairs in spinel ferrites with favorable distances and angles for magnetic interactions, are calculated with the equations given as in [14]. The interionic distances and interionic bond angles are tabulated in Table 4. The interionic bond angles (θ_1 – θ_5) values agrees well with that of [15].

3.4   UV – VISIBLE ANALYSIS

Energy of the incident photon (hv) can be related to the band gap (E_g) according to the Tauc relation [16, 17]

$$(\alpha hv)^2 = A(hv - E_g) \tag{7}$$

The value of the band gap energy can be calculated from the linear interpolation of the photon energy against $(\alpha hv)^2$ and it is estimated as 2.1eV. Nickel ferrite with band gap energy value of 2.5eV is reported by S.N.Dolia et.al [18].

3.5 MAGNETIC ANALYSIS

The magnetic data for the sample were recorded at room temperature using a vibrating sample magnetometer. The hysteresis loop, which shows the variation of magnetization as a function of an applied magnetic field is shown in Fig. 5. The maximum applied magnetic field is 20 kG. Saturation magnetization (M_s), coercivity (H_{ci}), retentivity (M_r), squareness ratio (M_r/M_s), Bohr magneton value (both calculated and observed) and the Yaffet-Kittel angle (θ_{YK}) of $NiFe_2O_4$ were calculated from the hysteresis loop and tabulated in Table 2.

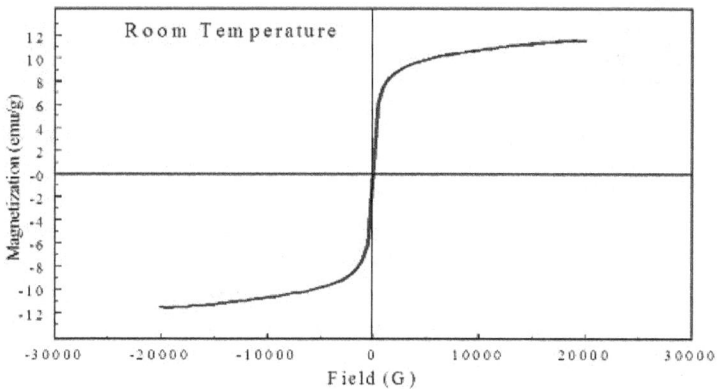

Figure 5. Hysteresis loop of $NiFe_2O_4$.

The 'S' shape of the hysteresis curve together with a small coercivity value reveals the presence of small magnetic particles exhibiting super paramagnetic behavior. The saturation magnetization (M_s) value of $NiFe_2O_4$ sample at 300 K is 11.6 emu/g. It is much lower when compared with that of bulk $NiFe_2O_4$ value, 55 emu/g [19] and this is attributed to (i) the core shell morphology of the nano particles consisting of ferrimagnetically aligned core spins and spin-glass-like surface layer [20] and (ii) the lower degree of crystalline of the sample [21]. The spin glass shell leads to decrease of the number of an aligned magnetic moment in the entire particle and in the present study this fact is reflected in the discrepancies between Bohr magneton calculated and observed value and the intensity of the (311) peak reflects the lower degree of crystallinity of the sample.

The Bohr Magneton value (saturation magnetization per formula unit in Bohr magneton at absolute temperature) evaluated from the hysteresis loop is given by

$$\mu_B^H \text{(Bohr Magneton)} = \left\{\frac{molecular\ weight}{5585}\right\} M_s \left(\frac{emu}{g}\right) \tag{8}$$

The value of Bohr Magneton calculated on the basis of the cation distribution and the Neel's two sublattice model, i.e. Neel's moment $\mu_B^N = M_B - M_A$ (where M_B and M_A are sublattice magnetizations) for the system is listed in Table 2. Due to the discrepancy between μ_B^H and μ_B^{Cal} values of the Bohr magneton, Neel two sub-lattice model cannot be used in this case. Hence, the Yaffet-Kittel (θ_{YK}) angle is calculated using the relation μ_B^H $= M_B \cos(\theta_{YK}) - M_A$. According to the Yaffet-Kittel model, the B sublattice can be split into two sublattices B_1 and B_2 having moments equal in magnitude and each making an angle (θ_{YK}) with the direction of the net magnetization at 0 K. This triangular spin arrangement of ions on the B site leads to the reduction in the A – B site exchange interaction, which is also confirmed from the MEM study, hence a low saturation magnetization value is observed.

3.6   DIELECTRIC ANALYSIS

The effect of frequency in the range of 0.1 Hz to 15 MHz, at room temperature on the dielectric constant, complex dielectric constant and dielectric loss tangent has been studied for the nickel ferrite sample calcined at 700 °C. The variation of dielectric constant (ε') and complex dielectric constant (ε'') at room temperature with frequency in logarithmic scale is shown in Fig. 6.

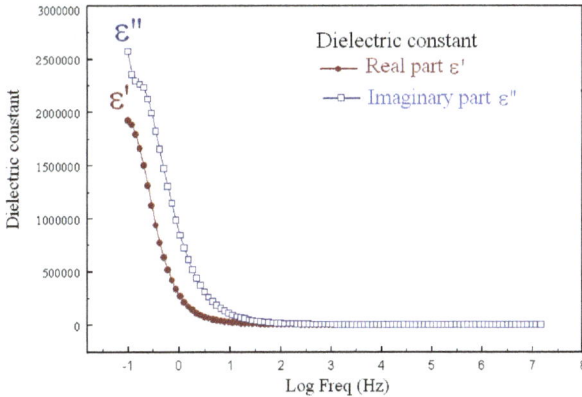

Figure 6. Effect of frequency on ε' and ε'' at room temperature for $NiFe_2O_4$.

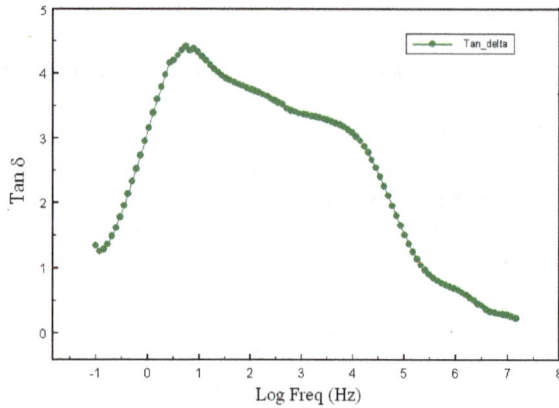

Figure 7. Plot of loss tangent against log frequency for $NiFe_2O_4$.

The fall in the dielectric constant value is steep in the lower frequency region whereas the fall is slower in the higher frequency region. Normal dielectric behaviour is observed since the value of the dielectric constant decreases with increase in frequency without showing any peak. Several investigators [22, 23] also reported normal dielectric behavior in ferrites. This normal dielectric behavior can be explained on the basis of Maxwell-Wagner interfacial polarization [24, 25] in agreement with Koops phenomenological theory [26]. In ferrites, the system may be thought of as consisting of well conducting grains separated by highly resistive grain boundaries. The charge carriers upon reaching the grain boundary by hopping encounter different resistance so that accumulation of charge at the separating boundaries occurs, causing interfacial polarization and hence the dielectric constant value is highly raised. As we can see a hopping electron in the direction of the external applied field increases the dielectric polarization, therefore, there is a strong correlation between the conduction mechanism and the dielectric behavior in ferrites. However as the frequency of the applied field is increased, the dielectric constant decreases until reaching a constant value, because beyond a certain frequency of the applied electric field the polarization cannot follow the alterations in the applied field [27]. In the present study, the dielectric constant value of 10^6 and 11 respectively is observed at the lowest frequency (0.1Hz) and highest frequency studied (15MHz). This high value (10^6) indicates that the available number of Fe^{2+} ions at the B site is higher. These ferrous ions on the B sites, which take part in the electron exchange interchange $Fe^{2+} \leftrightarrows Fe^{3+}$ are responsible for the polarization and hence dielectric constant in ferrites. Ferrites with high value (10^6) of dielectric constant are finds applications in microwave devices [28]. Circulators etc. [28]. Rangamohan et al [22] observed a value of around 32 for ε' at 100 kHz for nickel ferrite sample prepared by double sintering ceramic technique whereas a value of 34 is observed for ε' in this case at 100 kHz. From Fig. 6 it can be observed that both ε' and ε'' follow the same trend with increased frequency.

The plot of dielectric loss tangent (tanδ) against log frequency is shown in Fig. 7. It can be seen in the figure that initially the loss tangent increases and reaches a maximum and then decreases with the increase in frequency. In this investigated sample, tanδ shows a maximum at a frequency of 6 Hz. The occurrence of the peak in the tanδ versus log frequency curve for $NiFe_2O_4$ sample is explained [22] as follows. As pointed out by Iwauchi [29] there is a strong correlation between the conduction mechanism and the dielectric behavior of ferrites. The conduction mechanism in n-type ferrites is considered to be due to hopping of electrons between Fe^{2+} and Fe^{3+} and the hopping of holes between Ni^{3+} and Ni^{2+} in p-type ferrites. As such when the hopping frequency is nearly equal to that of the externally applied electric field, a maximum of loss tangent may be observed. The dielectric parameter (ε', ε'' and tanδ) values at selected frequencies are listed in Table 5.

Table 5 Dielectric parameters values of the Sample at 0.1Hz, 1kHz, 5kHz and 15MHz.

Parameter	Values				
	0.1 Hz	1kHz	0.1MHz	1 MHz	15MHz
ε'	1.92×10^6	6.4×10^2	34	17	11
ε''	2.57×10^6	2.1×10^3	51	11	2.4
tanδ	1.3	3.4	1.5	0.65	0.22

CONCLUSIONS

XRD data of the nickel ferrite sample prepared by using the solid state reaction method is refined using the Rietveld refinement method. The agreement between the theoretical and experimental lattice parameter values confirms the cation distribution. The experimental lattice parameter value is correlated with the number of Fe^{2+} ions at the octahedral site. Nano meter range particles sizes were confirmed from the morphology study. The XRF characterization shows the purity of the sample. The mid bond electron density values between various atoms at tetrahedral and octahedral sites are presented. The tetrahedral – tetrahedral site interaction is found to be strongest among the three interactions in the present study. The magnetic measurement reveals that the sample has low value of saturation magnetization. A high value (10^6) of dielectric constant is observed which is attributed to a higher number of Fe^{2+} ions at the octahedral site.

REFERENCES

[1]   Pengzhao Gao, Xia Hua, Volkan Degirmenci, David Rooney, Majeda Khraisheh, Robert Pollard, Robert M. Bowman, Evgeny V. Rebrov, Structural and magnetic properties of $Ni1\_xZnxFe2O4$ (x= 0, 0.5 and 1) nanopowders prepared by sol–gel method, J.Magn. Magn. Mater. 348 (2013) 44–50.
http://dx.doi.org/10.1016/j.jmmm.2013.07.060

[2]   F. Shahbaz Tehrani, V. Daadmehr, A.T. Rezakhani, R. Hosseini Akbarnejad S. Gholipour, Structural, Magnetic, and Optical Properties of Zinc- and Copper-Substituted Nickel Ferrite Nanocrystals, J Supercond Nov Magn DOI 10.1007/s10948-012-1655-5.

[3]   D.V. Kurmude R.S. Barkule A.V. Raut D.R. Shengule K.M. Jadhav, X-Ray Diffraction and Cation Distribution Studies in Zinc-Substituted Nickel Ferrite Nanoparticles, J Supercond Nov Magn DOI 0.1007/s10948-013-2305-2.

[4]   A. D. Sheikh, V. L. Mathe, Anomalous electrical properties of nanocrystalline Ni–Zn ferrite, J Mater Sci 43 (2008) 2018–2025.
http://dx.doi.org/10.1007/s10853-007-2302-6

[5]   Sagar E. Shirsath, B.G. Toksha, K.M. Jadhav, Structural and magnetic properties of In3+ substituted NiFe2O4, Mater. Chem.Phys. 117 (2009) 163–168.
http://dx.doi.org/10.1016/j.matchemphys.2009.05.027

[6]   S. M. Patange, S. E. Shirsath2, S.S. Jadhav, K. M. Jadhav, Cation distribution study of nanocrystalline $NiFe2\_xCrxO4$ ferrite by XRD, magnetization and Mossbauer spectroscopy, Phys. Status Solidi A 209, (2012) 347–352.
http://dx.doi.org/10.1002/pssa.201127232

[7]   H.M. Rietveld, A Profile Refinement Method for Nuclear and Magnetic, J. Appl. Crystallogr. 2 (1969) 65.
http://dx.doi.org/10.1107/S0021889869006558

[8]   V. Petricek, M. Dusek and L. Palatinus, (2006) Jana, The crystallographic computing system (Institute of Physics), Praha, Czech Republic.

[9]   Gull SF, Daniel GJ (1978) Image reconstruction from incomplete and noisy data. Nature 272 (1978) 686-690.
http://dx.doi.org/10.1038/272686a0

[10]   D.M. Collins DM (1982) Electron density images from imperfect data by iterative entropy maximization, Nature 49 298.

[11] A. D. Ruben, I. Fujio: Super-fast Program PRIMA for the Maximum-Entropy Method, Advanced materials Laboratory, National Institute for Materials Science, Ibaraki, Japan (2004), p. 305 0044.

[12] F. Izumi, R.A. Dilanian: Recent Research Developments in Physics, Part II, 3, Transworld, Research Network, Trivandrum, 2002, pp. 699–726.

[13] K. Momma, F. Izumi,VESTA: a three-dimensional visualization system for electronic and structural analysis J. Appl. Crystallogr. (2008) 41 653-658.
http://dx.doi.org/10.1107/S0021889808012016

[14] V. K. Lakhani, T. K. Pathak, N. H. Vasoya, K. B. Modi, Structural parameters and X-ray Debye temperature determination study on copper-ferrite-aluminates. Solid State Sci. 13 (2011) 539-547.
http://dx.doi.org/10.1016/j.solidstatesciences.2010.12.023

[15] T Slatineanu et al. Synthesis and characterization of nanocrystalline Zn ferrites substituted with Ni, Mater. Res. Bull. 46 (2011) 1455.
http://dx.doi.org/10.1016/j.materresbull.2011.05.002

[16] J.Tauc, R.Grigorovic, A.Vancu, physica status solidi (b), 15, 627-637. DOI:10.1002/pssb.19660150224.
http://dx.doi.org/10.1002/pssb.19660150224

[17] J.Pancove, optical process in semiconductors. Englewood Cliffs, NJ, USA: Prentice-Hall.

[18] S N Dolia, Rakesh Sharma, M P Sharma, N S Saxena, Synthesis, X-ray diffraction and optical band gap study of nanoparticles of $NiFe_2O_4$ Ind J Pure and Appl Phys 44, October 2006, 774-776.

[19] A. Goldman, Modern Ferrites Technology (New York Springer) 2006 Chap.2, p 32.

[20] G.Nabiyouni, M.Jafari Fesharaki, M.Mozafari, J.Amighian, Characterization and Magnetic Properties of Nickel Ferrite Nanoparticles prepared by ball milling technique, CHIN. PHYS.LETT., 27, (2010) 12640.
http://dx.doi.org/10.1088/0256-307X/27/12/126401

[21] J.Azadmanjiri, S.A.Seyyed Ebrahimi, H.K.Salehani, Magnetic properties of nanosize $NiFe_2O_4$ particles synthesized by sol–gel auto combustion method, Ceramics International 33 (2007) 1623-1625.
http://dx.doi.org/10.1016/j.ceramint.2006.05.007

[22] G.Rangamohan, D.Ravinder, A.V.Ramanareddy, B.S.Boyanov, Dielectric properties of polycryastalline mixed nickel-zinc ferrites, Mater.Letts, 40,(1999) 39-45.
http://dx.doi.org/10.1016/S0167-577X(99)00046-4

[23] U.N.Trivedi, M C Chhantbur K.B.Modi and H.H.Joshi, Frequency dependent dielectric behavior of cadmium and chromium co-sbstituted nickel ferrite, Ind. J Pure Appl.Phys. 43 september (2005) 688-690.

[24] J.C.Maxwell, Electricity and Magnetism, vol.1, Oxford univ.press, Oxford, 1929, section 328, p,752.

[25] K.W.Wagner, Anphys.(Leipzig) 40 (1913) 817.
http://dx.doi.org/10.1002/andp.19133450502

[26] C.G.Koops, Phys.Rev. 83 (1951) 121.
http://dx.doi.org/10.1103/PhysRev.83.121

[27] D.El.Kony, S.A.Saafan, Dielectric Properties and Magnetic Susceptibility of Mn-Zn Ferrites /SiO2 composites, J.Amer.Sci. 8 (10) 2012 51-57.

[28] A.V.Ramanareddy, G.Rangamohan, D.Ravinder, B.S.Boyanov, High- frequency dielectric behaviour of polycrystalline Zinc substituted cobalt ferrites, J Mater. Sci. 34 (1999) 3169-3176.
http://dx.doi.org/10.1023/A:1004625721864

[29] K.Iwauchi, Japan. J.Appl. Phys. 10, (1971) 1520.
http://dx.doi.org/10.1143/JJAP.10.1520

CHAPTER 12

Structural and Optical Properties of Li Doped Zirconia Nanoparticles

A.Abirami[1], M.Prema Rani[1]

[1] Research Centre and PG Department of Physics, The Madura College, Madurai-625 011,

Email: premaakumar@yahoo.com

Abstract

Li-stabilized cubic zirconia nanostructures ($Zr_{1-x}Li_xO_2$, x = 0.15, x = 0.25 and x = 0.35) have been prepared using a chemical precipitation method. The average size of the prepared crystallite was 17 nm. The electronic distributions in the unit cell were analyzed using the MEM method for the prepared cubic zirconia nanostructure. The bonding features were analyzed and it is found to behave like an ionic material. The energy band gap was determined via UV analysis and is found to be 4.13 eV for pure ZrO_2.

Keywords

Chemical Precipitation, XRD, Rietveld Refinement, Electron Density, Band Gap

Contents

1. INTRODUCTION

Zirconium oxide is a refractory material having high strength, high fracture toughness, excellent chemical resistance and low thermal conductivity. These properties of ZrO_2 make it highly useful in the field of structural, mechanical and high temperature applications [1, 2]. ZrO_2 is of great technological and scientific interest because of its potential applications in oxygen sensors, solid electrolytes, electrodes and fuel cells due to its oxygen ion conductivity and rich oxygen defects on its surface [3-7]. Pure zirconia exists in three crystallographic modifications; monoclinic, tetragonal and cubic phases [8-10]. The monoclinic form of zirconia is the thermodynamically stable phase at room temperature, whereas tetragonal and cubic are stable at high temperatures. Monoclinic zirconia transforms reversibly to the tetragonal phase when heated to about 1170 °C and as the temperature exceeds 2370 °C, the tetragonal phase transforms into the cubic phase. However, the high temperature phases may be stabilized at room temperature either by adding suitable dopants or by reducing the particle size into the nanometer regime [11]. The stabilization by the latter way (nano particles) has attracted considerable interest because of the large surface area, unusual adsorptive properties, surface defects, fast diffusivities and superplasticity of the nano-sized powder [12-13]. Addition of small percentage of some material such as Y, Ca, Mn, etc., results in the material that has superior thermal, mechanical and electrical properties. In this work Li doped zirconia nanostructures are synthesized and the electron density distribution and the bonding features are analysed.

2. EXPERIMENTAL

$Zr_{1-x}Li_xO_2$ (x = 0.15, 0.25, 0.35) nanomaterials have been prepared through chemical co-precipitating method [14]. 0.05 M of zirconyl nitrate solution was added with lithium acetate solution with a drop wise addition of tetra methyl ammonium hydroxide solution with vigorous stirring up to a ph level of eight. The resulting precipitate was washed and dried at 120 °C for 1 hour and finally the samples were sintered at 500 °C for 8 hours. The prepared samples were characterized by an X-ray diffractometer (XRD), a UV-Vis spectrometer and scanning electron microcopy for structural, optical and morphological studies respectively.

3. RESULTS AND DISCUSSION

3.1 X-RAY ANALYSIS

The X-ray characterization was done for the prepared powder samples of $Zr_{1-x}Li_xO_2$ (x = 0.15, 0.25, 0.35) nanomaterials using Bruker Eco D8 Advance at Kalasalingam

University, Madurai, using CuKα_1 radiation with a 2θ range of 5° to 120° and 0.02° step size. The observed X-ray peaks for the prepared nanostructures are matched with standard pattern from the Joint committee for Powder Diffraction Standards (JCPDS) XRD data set reported in the file (JCPDS No: 270997). Li stabilized cubic zirconia nanostructures are identified with the space group of Fm$\bar{3}$m. The effect of Li concentration on the structural features on ZrO$_2$ nanostructures has been analyzed further using the Rietveld refinement [15] method which is employed in the software JANA 2006 [16]. It has been used to fit the experimental and calculated diffraction patterns by considering the cubic structure with four molecules in the unit cell. The atomic coordinates for Zr and O atoms were set as (0, 0, 0) and (1/4; 1/4; 1/4) respectively. In the Rietveld refinement technique, structural parameters, lattice parameters, peak shift, background profile shape and preferred orientation are refined. The principle of the Rietveld Method is to minimize the difference between the theoretically modelled profile and the observed one. The Rietveld refined powder profiles of Zr$_{1-x}$Li$_x$O$_2$ (x = 0.15, x = 0.25, x = 0.35) nanostructures are shown in figures 1.The refined cell parameters are shown in table 1.

Table 1: Structural parameters of Zr$_{1-x}$Li$_x$O$_2$ nanostructures

Parameters	X=0	X=0.15	X=0.25	X=0.35
Cell Parameters a=b=c(Å)	5.0977(0.11)	5.0584(0.04)	5.0812(0.12)	5.1099(0.09)
Volume(Å^3)	132.47	129.43	131.19	133.43
Number of electrons in the unit cell	224	202	186	172
B-Debye Waller factor(Å^2)	2.4699	1.3983	0.7808	0.5988
Reliable index Robs (%)	1.77	2.06	1.83	1.91
Profile reliable index Rp (%)	5.61	4.88	5.18	5.06
Goodness of Fit	0.23	0.20	0.21	0.21

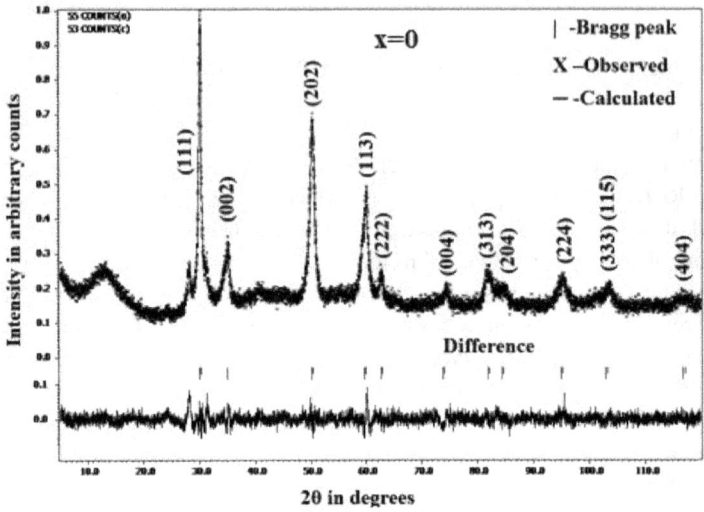

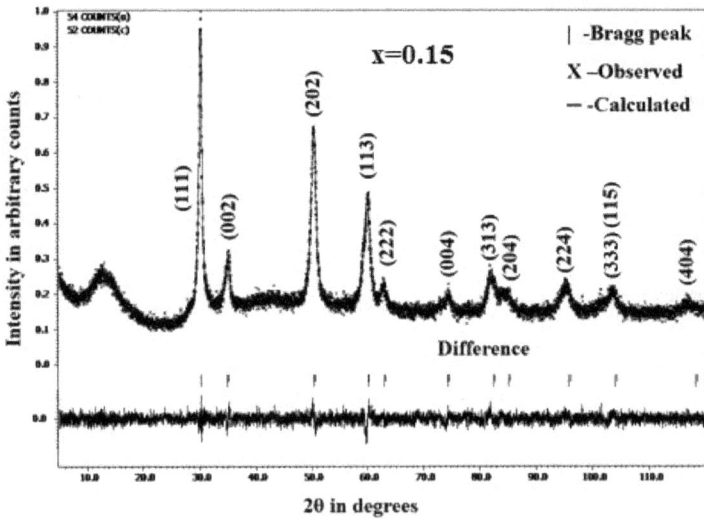

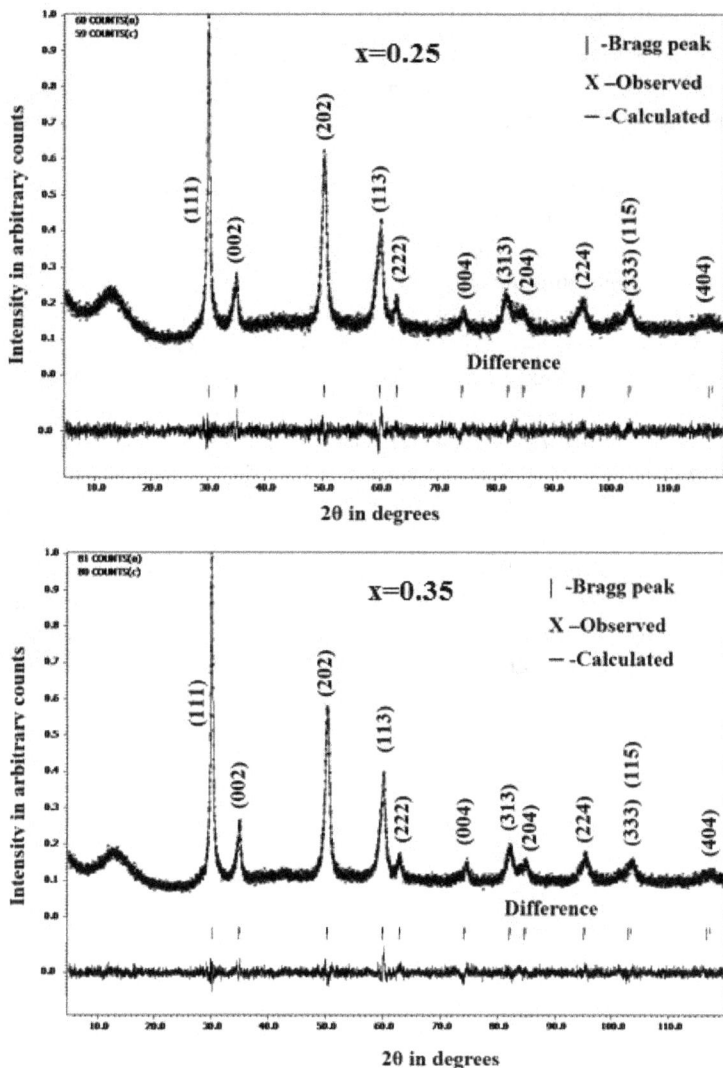

Fig. 1: Rietveld refined powder profiles of $Zr_{1-x}Li_xO_2$ (x = 0.15, x = 0.25, x = 0.35) nanostructures

3.2 CHARGE DENSITY ANALYSIS

The maximum entropy method (MEM) [17] is an important and accurate technique to observe the electron density distribution in the unit cell because of its probabilistic approach. Also, it only needs a minimum number of information from the observed XRD spectra and it yields the least biased information. The software used in this method is Practice Iterative MEM Analyses (PRIMA)[18]. The structure factors extracted from the Rietveld refinement technique were used for this study. The electron density distribution in the unit cell was constructed through the PRIMA software. The results are visualized by the visualization software VESTA (Visualization for Electronic and Structural Analysis) [19, 20]. Three dimensional electron density distributions of $Zr_{1-x}Li_xO_2$ nanostructures are shown in figure 2. The position of Zr and O atoms are clearly visible in the figure. The shaded region surrounded by the electron cloud shows the atoms Zr and O. Two dimensional electron density distributions for $Zr_{1-x}Li_xO_2$ nanostructures are shown in figure 3. This figure shows Zr and O atoms on the (110) miller plane. The variations in charge density distribution with varying dopant concentrations are clearly visualized in the two and three dimensional maps. The electron density distribution between the Zr and O atoms is shown in figure 4, as one dimensional electron density profile. The values of BCP (Bond Critical Point) between Zr and O atoms are given in table 2. For the sample $Zr_{0.85}Li_{0.15}O_2$, BCP between the Zr and the O atom has a value of 0.2709 $e/Å^3$. This low value attributed to the ionic bonding between the atoms. For the samples $Zr_{0.75}Li_{0.25}O_2$ and $Zr_{0.65}Li_{0.35}O_2$, the BCP value increases to 0.4209 $e/Å^3$ and 0.4381 $e/Å^3$ respectively. This shows an enhanced charge distribution but the magnitude of BCP is considerably lower and indicates the ionic nature of the samples.

X=0

X=0.15

X=0.25

X=0.35

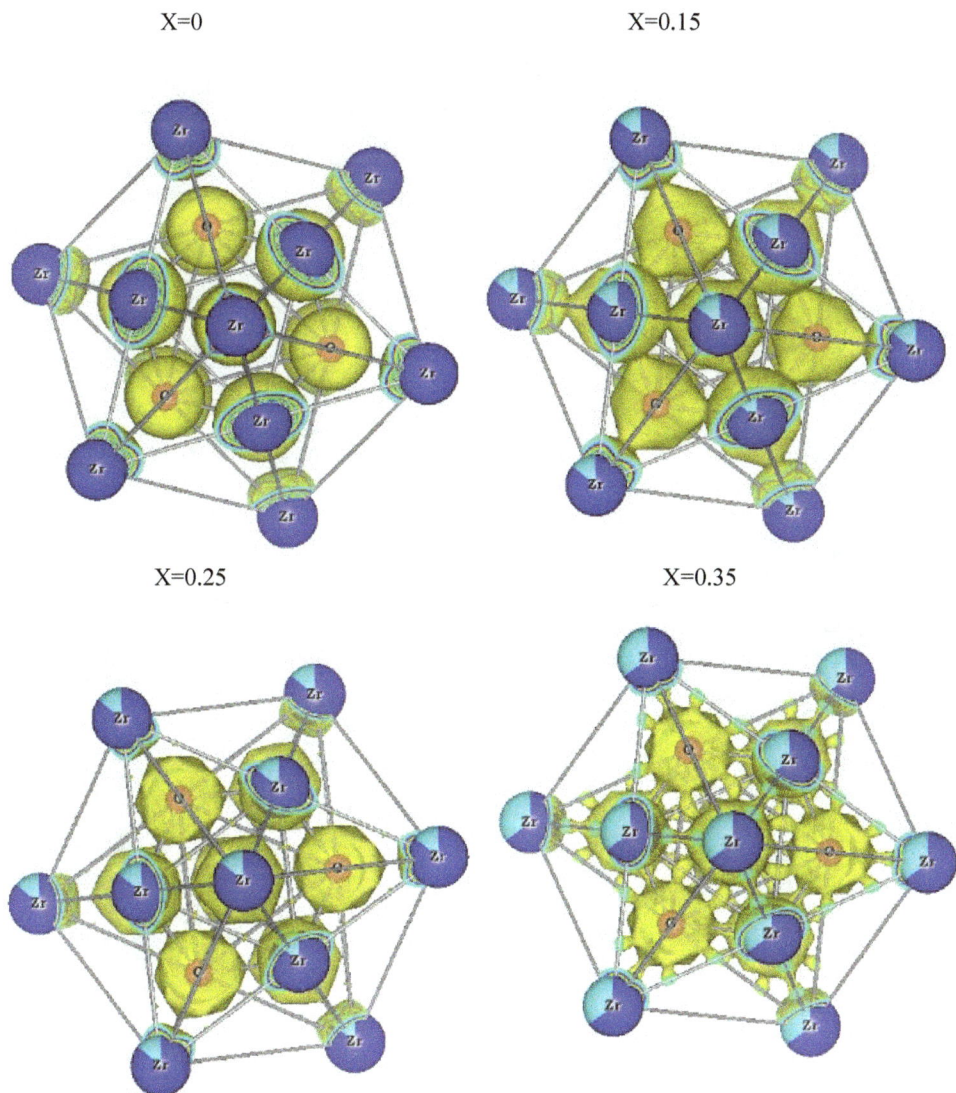

Fig. 2 Three dimensional electron density distribution in the unit cell (isosurface level = $0.6 \, e/\text{Å}^3$) for $Zr_{1-x}Li_xO_2$ nanostructures X=0, X = 0.15, X = 0.25 and X = 0.35

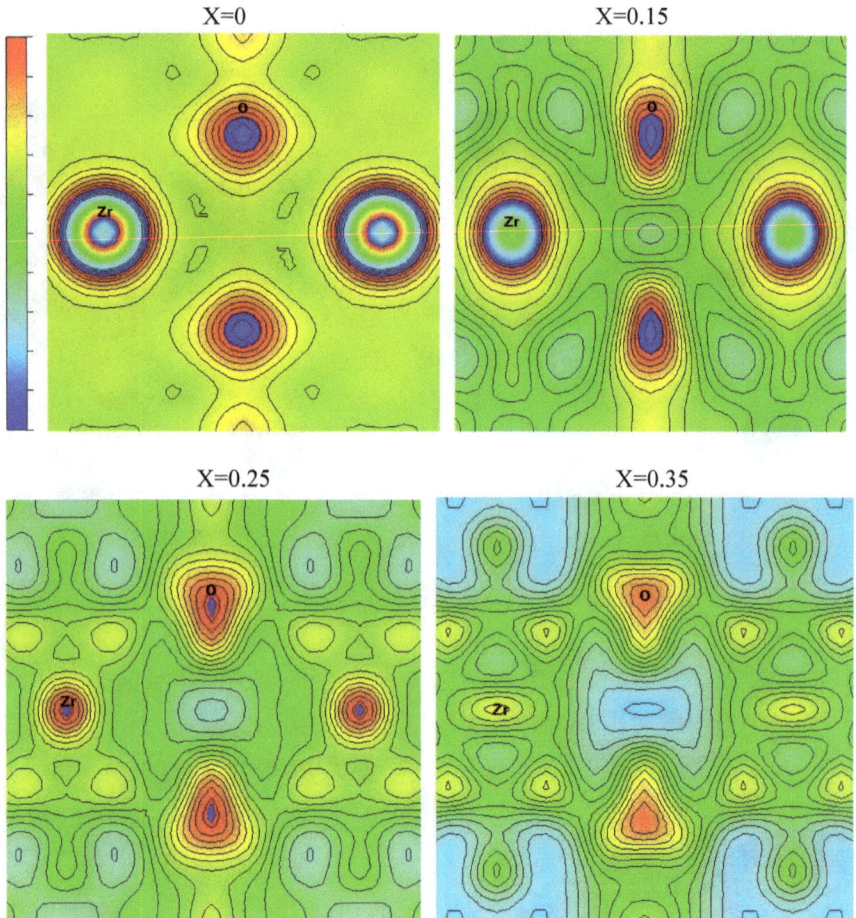

Fig. 3 Two dimensional electron density distribution in the unit cell in the contour range of 0 to 1.0 e/$Å^3$ with contour interval of 0.05 e/$Å^3$ for $Zr_{1-x}Li_xO_2$ nanostructures X=0, X = 0.15, X = 0.25 and X = 0.35

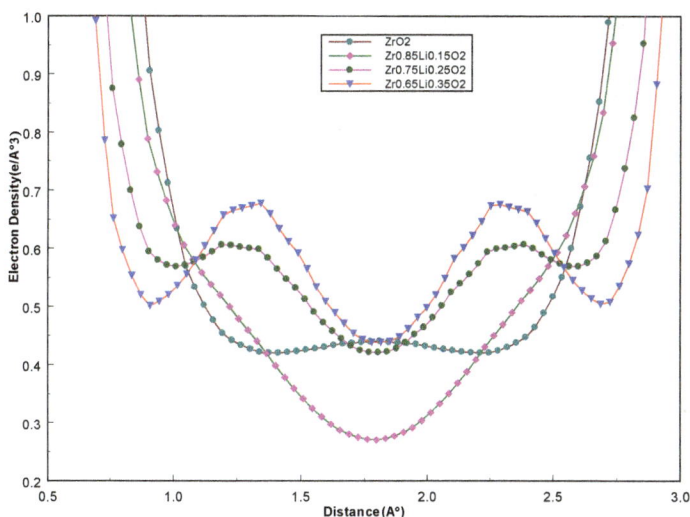

Fig. 4 One dimensional electron density profiles for Zr1-xLixO2 nanostructures between Zr and Oxygen atoms

Table2: BCP (Bond Critical Point) at Zr–O bond of Zr$_{1-x}$Li$_x$O$_2$ nanostructures

Sample	Position (Å)	Electron density (e/Å^3)
X=0	1.7389	0.4382
X=0.15	1.7615	0.2709
X=0.25	1.7694	0.4209
X=0.35	1.7794	0.4381

3.3 ANALYSIS FROM ULTRAVIOLET SPECROSCOPY

The UV analysis of the sample was done at SAIF (Sophisticated Analytical Instrument Facility), Cochin, in the wavelength range of 200 to 800 nm. The UV spectra are shown in figure 5 for Zr$_{1-x}$Li$_x$O$_2$ nanostructures. From the UV plot, the energy band gap is measured with the help of Tauc relation, the graph is plotted between (hv) vs (αhv)^2(eV/cm)^2. Where, α is the Absorption coefficient, h is the Planck's constant. v is the frequency, and E$_g$ is the energy band gap. Band gap of Zr$_{1-x}$Li$_x$O$_2$ nanostructures are tabulated in table 3. The energy band gap of ZrO_2 was determined as 4.13 eV. The band gap varies with doping. Band gap for $Zr_{0.85}Li_{0.15}O_2$ is larger compared to other doping concentrations. This is also observed from MEM calculations in table 2.

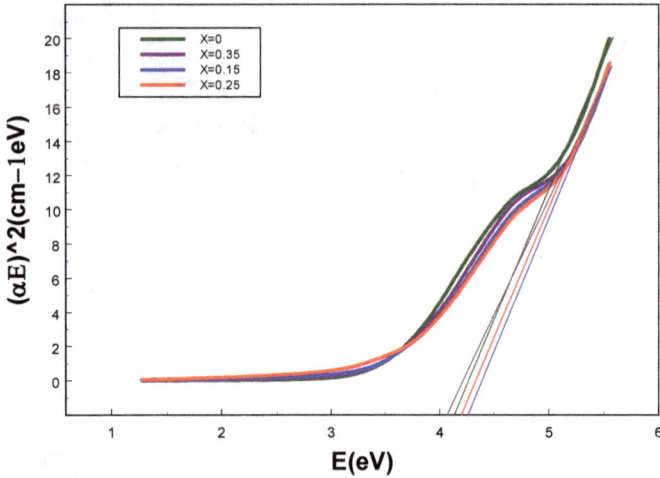

Fig. 5 UV-vis absorption spectra of Zr1-xLixO2 nanostructures

Table 3: Determination of bandgap from UV

Sample	Band gap(eV)
ZrO_2	4.13
$Zr_{0.85}Li_{0.15}O_2$	4.40
$Zr_{0.75}Li_{0.25}O_2$	4.23
$Zr_{0.65}Li_{0.35}O_2$	4.07

3.4 SEM DATA DETAILS

SEM analysis has been done at SAIF (Sophisticated Analytical Instrument Facility), Cochin. The topographical information of ZrO_2 and Lithium doped ZrO_2 is given in figure 6. Average crystallite size from the SEM picture is about 250 to 300 nm.

Fig. 6 Morphological image (SEM) of the prepared Zr1-xLixO2 nanostructures

CONCLUSIONS

Li-doped ZrO_2 nanostructures were synthesized by the precipitation method. The average crystallite size was determined, using the Scherrer formula, as 17 nm. The cell parameters decrease for $Zr_{0.85}Li_{0.15}O_2$, compared to other concentrations. The charge density distributions in the unit cell were analyzed using the MEM method. From the quantitative values of the charge distribution the bonding in the samples is observed to be ionic. The band gap was determined from UV analysis, and the band gap for ZrO_2 was determined as 4.13 eV. An increase in bandgap is observed for $Zr_{0.85}Li_{0.15}O_2$.

REFERENCES

[1]   B. Basu, Toughening of Y-stabilized tetragonal zirconia ceramics, Int. Mater. Rev. 50 (2005) 239–256.
       http://dx.doi.org/10.1179/174328005X41113

[2]   I. Berkby, R. Steven, Applications of zirconia ceramics, Key. Eng. Mater. 527 (1996) 122-124.
       http://dx.doi.org/10.4028/www.scientific.net/kem.122-124.527

[3]   M. Maczka, E.T.G. Lutz, H.J. Verbeek, K. Oskam, A. Meijerink, J. Hanuza, M. Stuivinga, Spectroscopic studies of dynamically compacted monoclinic ZrO2, J. Phys. Chem. Solids 60 (1999) 1909–1914.
       http://dx.doi.org/10.1016/S0022-3697(99)00221-8

[4]   Andreas Ziehfreund, Ulrich Simon, Wilhelm F. Maier, Oxygen ion conductivity of platinum-impregnated stabilized zirconia in bulk and microporous materials, Adv. Mater. 8 (1996) 424-427.
       http://dx.doi.org/10.1002/adma.19960080512

[5]   F.P.F. van Berkel, F.H. van Heuveln, J.P.P. Huijsmans, Characterization of solid oxide fuel cell electrodes by impedance spectroscopy and I–V characteristics, Solid State Ionics 72 (1994) 240-247.
       http://dx.doi.org/10.1016/0167-2738(94)90153-8

[6]   Henk Verweij, Nanocrystalline and Nanoporous Ceramics, Adv. Mater. 10 (1998) 1483-1486.
       http://dx.doi.org/10.1002/(SICI)1521-4095(199812)10:17<1483::AID-ADMA1483>3.0.CO;2-J

[7]   Marc Mamak, Neil Coombs, Geoffrey Ozin, Self-Assembling Solid Oxide Fuel Cell Materials: Mesoporous Yttria-Zirconia and Metal-Yttria-Zirconia Solid Solutions, J. Am. Chem. Soc, 122 (2000) 8932-8939.
       http://dx.doi.org/10.1021/ja0013677

[8]   E.H.Kisi,C.J. Howard, Crystal Structures of Zirconia Phases and their Inter-Relation, Key. Eng. Mater 153(1998) 1–36.
       http://dx.doi.org/10.4028/www.scientific.net/KEM.153-154.1

[9]   Shukla S, Seal S, Synthesis of Tetragonal Phase Stabilized Nano and Submicron
       Sized Nanoparticles, U.Int. Mater.Rev. 50 (2005) 45-64.
       http://dx.doi.org/10.1179/174328005X14267

[10]   A.E. Bohe, J. Andrade-Gamboa, D.M. Pasquevich, A. Tolley,J.L. Pelegrina, Size-
       dependent density of zirconia nanoparticles, J. Am. Ceram. Soc. 83(2000) 755-
       760.

[11]   R.C. Garvie., The Occurrence of Metastable Tetragonal Zirconia as a Crystallite
       Size Effect. J. Phys. Chem. 69 (1965) 1238-1243.
       http://dx.doi.org/10.1021/j100888a024

[12]   H.D. Gesser, P.C. Goswami, - Aerogels and related porous materials, Chem. Rev.
       89 (1989) 765-788.
       http://dx.doi.org/10.1021/cr00094a003

[13]   C. Suryanarayana, C.C. Koch, Nanocrystalline materials-Current research and
       future directions, Hyperfine Interactions 130 (2000) 5-44.
       http://dx.doi.org/10.1023/A:1011026900989

[14]   V.R. Choudhary, S. Banerjee, S.G. Pataskar, Low temperature complete
       combustion of dilute propane over Mn-doped ZrO_2 (cubic) catalysts, Journal of
       chemical sciences, 115 (2003) 287-298.
       http://dx.doi.org/10.1007/BF02704220

[15]   H.M. Rietveld, A profile refinement method for nuclear and magnetic structures, J.
       Appl. Crystallogr. 2 (1969) 65-71.
       http://dx.doi.org/10.1107/S0021889869006558

[16]   V. Petrˇicˇek, M. Dusˇek, L. Palatinus, in: JANA2000, The Crystallographic
       Computing System, Institute of Physics, Academy of Sciences of the Czech
       Republic, Praha, 2000.

[17]   S. Kumazawa, Y. Kubota, M. Takata, M. Sakata, Y. Ishibashi, MEED: a program
       package for electron-density-distribution calculation by the maximum-entropy
       method, J. Appl. Crystallogr. 26(1993) 453-457.
       http://dx.doi.org/10.1107/S0021889892012883

[18]   A. D. Ruben, I. Fujio: Super-fast Program PRIMA for the Maximum-Entropy
       Method, Advanced materials Laboratory, National Institute for Materials Science,
       Ibaraki, Japan p. 305 (2004) 0044.

[19]   F.Izumi, R.A. Dilanian: Recent Research Developments in Physics, Part II, 3,
       Transworld, Research Network, Trivandrum, 2002, pp. 699–726.

[20] K. Momma, F. Izumi, VESTA: a three-dimensional visualization system for electronic and structural analysis, J. Appl. Crystallogr. 41 (2008) 653. *http://dx.doi.org/10.1107/S0021889808012016*

CHAPTER 13

Synthesis, Structure and Magnetic Behavior of Ce-Doped Lanthanum Manganite Ceramics

N. Thenmozhi[1], R. Saravanan[2]

[1]PG and Research Department of Physics, NMSSVN College, Nagamalai, Madurai-625 019, India

[2]Research Centre and Post Graduate Department of physics, The Madura College, Madurai – 625 011, India

Email:saragow@gmail.com; thenmozhi.n6@gmail.com

Abstract

The ceramic perovskite $La_{0.88}Ce_{0.12}MnO_3$ has been synthesized using the solid state reaction method. The grown sample was characterized for its structural, optical, and magnetic properties using powder X-ray diffraction, UV–visible spectra and vibrating sample magnetometer (VSM) measurements. The Rietveld analysis of the X-Ray diffraction (XRD) profile clearly indicated that the XRD pattern is well fitted with an orthorhombic structure. The electronic structure of the synthesized sample has been studied through the maximum entropy method. The optical study gives the direct band energy gap value as 2.106 eV. VSM measurements, at room temperature show paramagnetic behavior of the prepared sample.

Keywords

Perovskite, X-ray diffraction, electronic structure, bandgap, magnetic properties

Contents

1.   INTRODUCTION

The rare-earth manganites were studied over the past fifty years due to their remarkable magnetic behavior. Generally, rare-earth manganites are in the form of cubic perovskite structure of ABO_3 type. Lanthanum manganite ($LaMnO_3$) is the parent compound of rare earth manganites which have the general chemical formula $R_{1-x}A_xMnO_3$, where R=La^{3+} and A= Sr^{2+}, Ca^{2+}, Ba^{2+}, etc. The La ion occupies the corner site of the cube; the Mn ion occupies the center of the cubic structure and the oxygen ions surround the manganese ion in the form of an octahedron [1]. The Goldschmidt tolerance factor which is a measure of the deviations from the ideal cubic perovskite structure is t= (R_A+R_B) /$\sqrt{2}(R_B+R_O)$, where R_A, R_B and R_O are the averaged ionic radii of A, B and the oxygen ion respectively. The deviation from the tolerance factor t=1 gives the distorted perovskite structure in the form of orthorhombic and rhombohedral structures. The crystal structure of LaMnO3 was found to vary from orthorhombic (Pbnm) to rhombohedral (R-3c), depending on the method of synthesis [2]. In $LaMnO_3$, the hole-doping could be achieved by the substitution of divalent cations such as Ba, Ca, Sr, Pb etc. in the La-site. However, the electron doped compounds could be achieved in $LaMnO_3$ by substituting Ce^{4+} ions in La-site [3] and this work was done first by Mandal and Das [4]. By the substitution of Ce^{4+} on La^{3+}, they observed similar behaviors of hole-doped manganites. One important observation from previous studies is that Ce doped $LaMnO_3$ bulk samples of $La_{1-x}Ce_xMnO_3$ cannot form in a single phase. But using the pulsed laser deposition technique, there exist Ce doped $LaMnO_3$ in a single phase [5]. The Ce doped bulk $LaMnO_3$ is a multiphase mixture of lanthanum deficient manganate phases and Ce oxides. Studies like Hall effects and magneto resistance effects were performed on Ce doped $LaMnO_3$ and it seems that all the data obtained are consistent with each other. The other tetravalent ion-doped lanthanum manganites $La_{1-x}A_xMnO_3$, A=Sn, Zr, Sb and Te were also studied and the same behavior as Ce doped $LaMnO_3$ [6] was observed. Lanthanum manganite ceramic materials have a wide variety of applications because of their ionic conduction, magnetic, thermal and mechanical properties etc. For example, the ionic conduction property of these materials is useful in oxygen sensors and solid oxide fuel cells [7]. These materials are also used as catalysts, semiconductor ceramic materials, ferromagnets and heating elements [8]. The doping of divalent elements like Ca, Sr, and Ba on the La-site of lanthanum manganite, leads to a paramagnetic to ferromagnetic transition, which is applicable to reading/recording heads, magnetic sensors and magnetic refrigerators [9]. In the present work, the La site is doped with cerium and the effect of Ce doping on structural, optical and magnetic properties has been investigated.

2. EXPERIMENTAL

2.1 SAMPLE PREPARATION

The polycrystalline sample of $La_{0.88}Ce_{0.12}MnO_3$ was prepared using the solid state reaction technique. Stoichiometric amounts of La_2O_3, CeO_2 and MnO_2 with a purity of 99.99% were mixed and ground for about 2 h. Initially, the La_2O_3 powder was preheated at 1000 °C for 12 h. The powder sample was then calcined at 1050 °C for 15 h. Then the reacted powder was reground for 1 h and then pelletized. Finally, these pellets were annealed at 1350 °C for 15 h.

2.2 CHARACTERIZATIONS

The synthesized sample has been analyzed using powder X-ray diffraction. The XRD datasets were collected at Sophisticated Analytical Instrument Facility (SAIF), Cochin University, Cochin, using an X-ray diffractometer (Bruker AXS D8 advance) with CuKα, monochromatic incident beam (λ=1.54056 Å) with a step size of 0.02°. The UV-visible absorption spectrum for the grown sample was recorded at SAIF, Cochin (India) using an UV-visible spectrometer Cary 5000 (Varian, Germany). The magnetic properties of the grown sample were investigated using a vibrating sample magnetometer (Lakeshore VSM 7410) measurements which were also taken at SAIF, IIT Madras, Chennai.

3. RESULTS AND DISCUSSION

3.1 POWDER X-RAY DIFFRACTION ANALYSIS

The untreated raw X-ray diffracted profile of the prepared sample is shown in figure 1. The observed XRD pattern matched well with that of the orthorhombic space group (No.62). An additional peak due to CeO_2 is noted in figure 1, which was also observed by Mitra et. al [10]. The Rietveld method [11] is used to refine the crystal structure model of a material. It can be used for quantitative phase identifications; lattice parameter and crystallite size calculations and determination of atom positions and occupancies. In the present work, to analyze the structural parameters of the samples, the collected XRD data were subjected to the Rietveld method [11] of refinement using Jana 2008 [12]. For the grown samples, cell parameters, peak shift, background profile shape and preferred orientation were refined from the observed XRD profiles by comparing them with the theoretically generated profiles. The crystal structure of the grown sample $La_{0.88}Ce_{0.12}MnO_3$ was refined in the orthorhombic setting which belongs to the space group of Pnma with four molecules in the unit cell. The refined profile for the grown sample $La_{0.88}Ce_{0.12}MnO_3$ is shown in figure 2. In the orthorhombic setting, La and Ce atoms are fixed at the Wyckoff position 4C (0.0267, 0.25, -0.004); the Mn atom is fixed at 4b (0, 0, 0.25); the O1 apex atom is at 4C (0.4905, 0.25, 0.0684) and the O2 planar atom at 8d (0.2193, 0.5361, 0.2195). The refined structural parameters and the reliability indices for the prepared sample are given in table 1. The lattice parameters and hence the volume of the unit cell is decreased for the grown sample, compared with undoped $LaMnO_3$ [13]. The reason for this may be due to the ionic radius of Ce^{4+} (0.97Å) being smaller than the ionic radius of La^{3+} (1.16Å) [14].

The average grain size of the synthesized sample $La_{0.88}Ce_{0.12}MnO_3$ was estimated from XRD powder data using GRAIN software [15] which employs Scherrer formula [16] $t=0.9\lambda/\beta\cos\theta$, where t is the grain size (size of the coherently diffracting domain), λ is the wavelength of X-ray, β is the full width at half maximum and θ is the Bragg angle of reflection. The estimated grain size of $La_{0.88}Ce_{0.12}MnO_3$ by the Scherrer equation is about 23 nm which agrees with the previously reported values [3, 17].

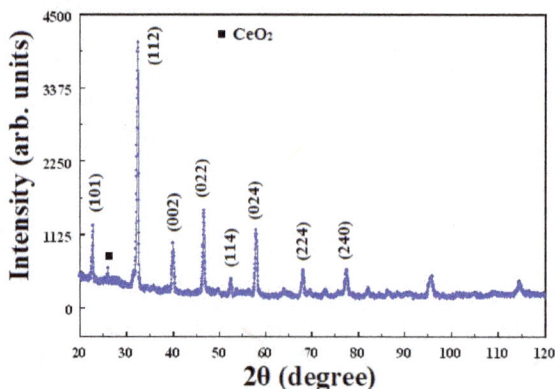

Figure 1. Observed X-ray powder diffractogram of $La_{0.88}Ce_{0.12}MnO_3$.

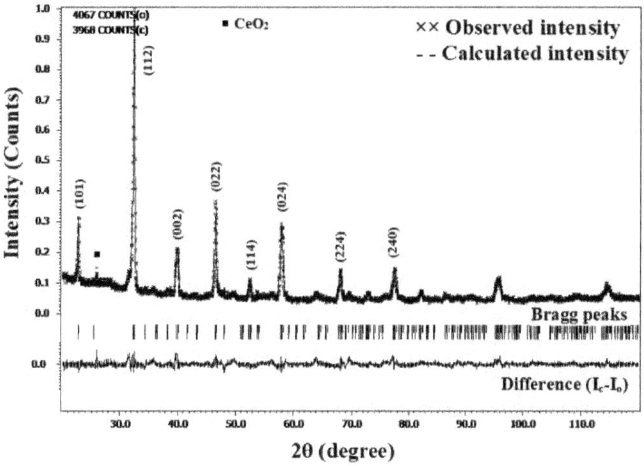

Novel Ceramic Materials

Figure 2. Fitted powder XRD profile for La$_{0.88}$Ce$_{0.12}$MnO$_3$.

Table 1. Structural parameters for La$_{0.88}$Ce$_{0.12}$MnO$_3$ through refinement of powder XRD data.

Parameters	Values
a (Å)	5.5563(10)
b (Å)	7.8209(12)
c (Å)	5.5150(10)
α=β=γ (°)	90
Volume (Å^3)	239.66 (11)
Density (gm/cc)	6.70(1)
R$_p$ (%)	7.12
R$_{obs}$ (%)	6.24
GOF	1.56
F(000)	424

181

3.2 CHARGE DENSITY ANALYSIS

The electron density distribution in the unit cell has been studied by the Maximum Entropy Method (MEM) [18] which is an accurate technique because of their statistical approach. This method uses structure factors extracted from the Rietveld refinement technique (11) and gives accurate pictures of the distribution of charges especially in the valence region. Hence, the MEM method is used for analyzing the bonding features and other structure based parameters. This method was implemented by the software called PRIMA (Practice Iterative MEM Analysis) [19] which is a FORTRAN program to determine the electron density by MEM [18] from X-ray data. The MEM computations for the grown sample $La_{0.88}Ce_{0.12}MnO_3$ was carried out using 48x64x48 pixels along the a, b and c axes of the orthorhombic lattice. The resultant electron density distribution in the unit cell was visualized using the visualization software VESTA (Visualization for Electronic and Structural Analysis) [20]. The 3-dimensional electron density distributions of $La_{0.88}Ce_{0.12}MnO_3$ are shown in figure 3 by considering the iso-surface level of 3.0 $e/Å^3$. The three dimensional electron density distribution of $La_{0.88}Ce_{0.12}MnO_3$ clearly gives the position of the atoms and confirms the orthorhombic phase of the grown sample. Figure 4 (a) shows the 3D unit cell with the (020) plane shaded. The two dimensional electron density distribution of Mn and O2 atoms are drawn in the range of 0-1.0 $e/Å^3$ with an interval of 0.04 $e/Å^3$ for the (020) plane and is shown in figure 4 (b). The 2-D map for the plane (020) (figure 4(b)) showing the increase of charges in the bonding region of Mn and O2 atoms, confirms that the bond Mn-O2 is covalent in character. The one dimensional electron density line profiles for the two bonds Mn-O1 and Mn-O2 are presented in figures 5 and 6. The mid bond density values and bond lengths corresponding to Mn-O1 and Mn-O2 bonds are given in table 2. The mid bond density values of both Mn-O1 and Mn-O2 bonds show that, both the bonds are covalent in nature. The bond length values for Mn-O1 and Mn-O2 agree well with the reported values [1, 21].

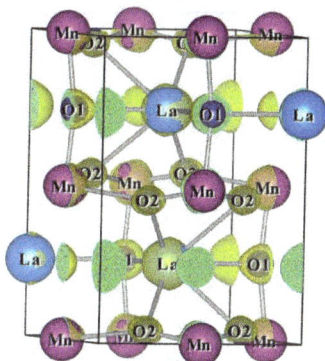

Figure 3. Three dimensional electron density isosurfaces for La$_{0.88}$Ce$_{0.12}$MnO$_3$.

(a)   (b)

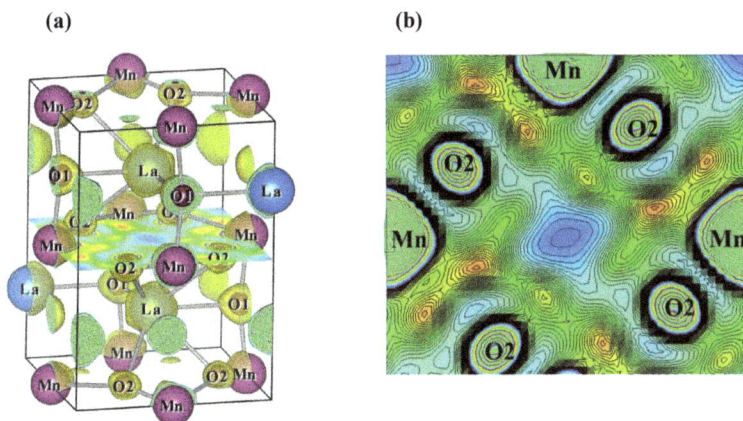

Figure 4. a) 3D unit cell of La$_{0.88}$Ce$_{0.12}$MnO$_3$ with (020) plane shaded. b) Two dimensional electron density distribution on (020) plane for La$_{0.88}$Ce$_{0.12}$MnO$_3$.

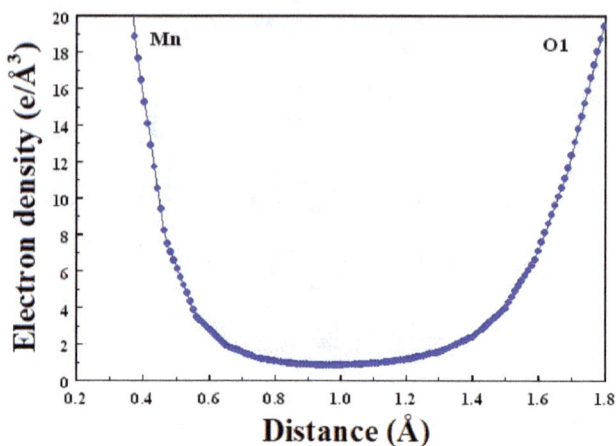

Figure 5. One dimensional electron density profile along Mn and O1 atoms in $La_{0.88}Ce_{0.12}MnO_3$.

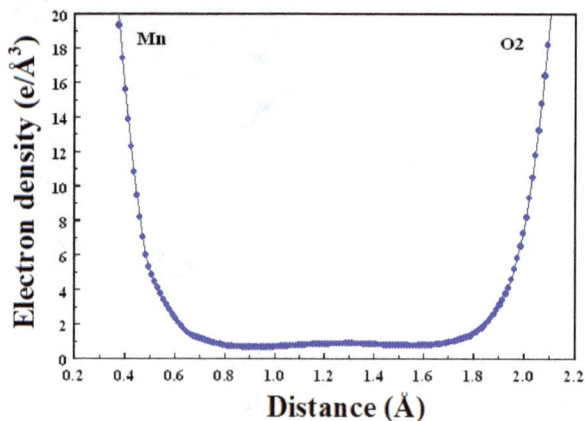

Figure 6. One dimensional electron density profile along Mn and O2 atoms in $La_{0.88}Ce_{0.12}MnO_3$.

Table 2 Bond lengths and mid bond electron densities for Mn-O1 and Mn-O2 bonds for $La_{0.88}Ce_{0.12}MnO_3$.

Mn-O1		Mn-O2	
Bond length (Å)	Mid bond electron density (e/Å³)	Bond length (Å)	Mid bond electron density (e/Å³)
1.9647	0.8516	2.3541	0.8319

3.3 OPTICAL PROPERTIES

The UV-Visible absorption spectrum for the grown sample was taken to estimate the band gap of the material. The UV absorption spectrum was taken in the range of 200 nm to 2000 nm. The band gap was estimated by the Tauc's relation [22] $\alpha h\gamma = A (h\gamma - E_g)^n$, where A is a constant, α is the absorption coefficient, $h\gamma$ is the photon energy, E_g is the energy band gap, n=1/2 for direct band gap materials and n=2 for indirect band gap materials. Using the Tauc's relation, a graph is drawn with the energy value in X-axis and $(\alpha h\gamma)^2$ in Y-axis and extrapolating the linear portion of the curve as shown in figure 7 gives the estimate of the direct band gap. The direct energy band gap of the synthesized sample from the Tauc's plot is 2.1057eV.

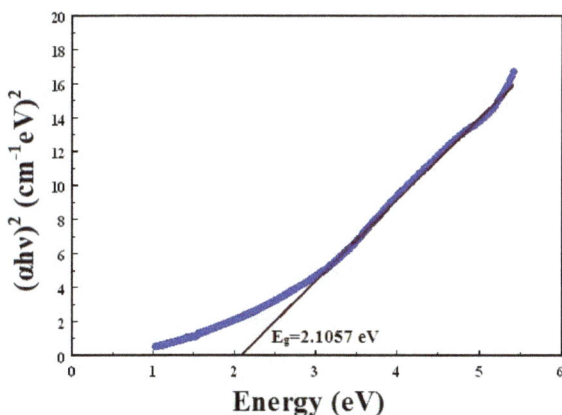

Figure 7. UV-Visible plot for $La_{0.88}Ce_{0.12}MnO_3$.

3.4   MAGNETIC STUDIES

$LaMnO_3$ is an antiferromagnetic insulating material below the Neel temperature $(T_N \sim 140K)$ and it behaves as a paramagnetic insulating material above T_N [23]. The ferromagnetic property and the metal-insulator transition (MIT) property of the electron doped $LaMnO_3$ was explained by Zener double exchange between Mn^{3+} and Mn^{2+} ions. The transition from paramagnetic to ferromagnetic behavior at just below T_C of the electron doped $LaMnO_3$ was explained by a polaron effect due to a strong electron-phonon coupling arising from the Jahn-Teller distorsion of the Mn^{3+} ions [10]. The M-H plot at T=300 K for the Ce doped $LaMnO_3$ sample is shown in figure 8 and the magnetic parameters from the VSM measurements are presented in table 3. The sample shows paramagnetic behavior as it is showing in the linear M Vs H curve [24].

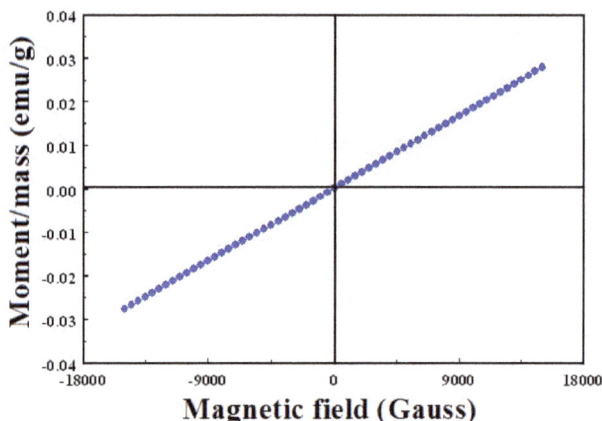

Figure 8. M-H curve for $La_{0.88}Ce_{0.12}MnO_3$ obtained by vibrating sample magnetometry measurements.

Table 3. Magnetic parameters from VSM measurements for $La_{0.88}Ce_{0.12}MnO_3$.

Parameters	Values
$M_s \times 10^{-3}$ (emu g^{-1})	27.739
H_c (G)	1.4060
$M_r \times 10^{-3}$ (emu g^{-1})	2.6720

CONCLUSION

The $La_{0.88}Ce_{0.12}MnO_3$ perovskite was synthesized by the high temperature solid state reaction technique. The structure analysis reveals that $La_{0.8}Ce_{0.12}MnO_3$ has orthorhombic symmetry with a space group Pnma. With the incorporation of Ce, it is found, the unit cell volume decreases compared with Ce free $LaMnO_3$. The average grain size for the sample is found to be 23 nm. MEM analysis reveals that Mn-O bonds are covalent in nature. M-H curve at T=300 K for $La_{0.8}Ce_{0.12}MnO_3$ sample shows paramagnetic behavior.

REFERENCES

[1] Wong Jen Kuen, Lim Kean Pah, Abdul Halim Shaari, Chen Soo Kien, Ng SiauWei and Albert Gan Han Ming, Effect of Rare Earth Elements Substitution in La site forLaMnO3 Manganites, Pertanika J. Sci. & Technol. 20 (1): (2012) 81 – 88

[2] Mats Johnsson and Peter Lemmens, Crystallography and Chemistry of Perovskites, Handbook of Magnetism and Advanced Magnetic Materials. Edited by Helmut Kronmuller and Stuart Parkin. Volume 4: Novel Materials.2007 John Wiley & Sons, Ltd. ISBN: 978-0-470-02217-7

[3] Kavita Bajaj, Vivas Bagwe, John Jesudasan, Pratap Raychaudhuri, Bandwidth control effects in electron doped manganiteLa0.7KxYxCe0.3MnO3 thin films, Solid State Communications, xx (2006) 1–4

[4] P. Mandal and S. Das, Transport properties of Ce-doped RMnO3 (R=La, Pr, and Nd) manganites, Phys. Rev. B 56, (1997) 15073
http://dx.doi.org/10.1103/PhysRevB.56.15073

[5] S. Das, A. Poddar. B. Roy, S. Giri, Studies of transport and magnetic properties of Ce-doped $LaMnO_3$, Journal of Alloys and Compounds, 365 (2004) 94-101
http://dx.doi.org/10.1016/S0925-8388(03)00688-1

[6] D.J. Wang, J.R. Sun, S.Y. Zhang, G.J. Liu, and B.G. Shen, Hall effect in $La_{0.7}Ce_{0.3}MnO_{3+\delta}$ films with variable oxygen content, Phys. Rev. B 73, (2006) 144403
http://dx.doi.org/10.1103/PhysRevB.73.144403

[7] Q. Shu, J. Zhang, B. Yan, J. Liu, Phase formation mechanism and kinetics in solid state synthesis of undoped and calcium-doped lanthanum manganite. Materials Research Bulletin, 44 (2009) 649-653
http://dx.doi.org/10.1016/j.materresbull.2008.06.022

[8] Nikolina L. Petrova, Dimitar S. Todorovsky, Veselinka G. Vasileva, Synthesis and characterization of Mn-, La-Mn- and La-Ca-Mn-citrates as precursors for $LaMnO_3$ and $La_{1-x}Ca_xMnO_3$, Central European Journal of Chemistry 3(2) (2005) 263–278

[9] Chul-min Heo, Min-sook Lee and Seong-Cho Yu, Magnetocaloric Effect of Perovskite Manganites of $La_{0.8}A_{0.2}MnO_3$ (A = Ca, Sr, Ba), Journal of the Korean Physical Society, 57 (2010) 1893-1896
     http://dx.doi.org/10.3938/jkps.57.1893

[10] C. Mitra, P. Raychaudhuri, J. John, S.K. Dhar, A.K. Nigam, and R Pinto, Growth of epitaxial and polycrystalline thin films of the electron doped system $La_{1-x}Ce_xMnO_3$ through pulsed laser deposition, J. Appl. Phys., Vol. 89, (2001) 524-530
     http://dx.doi.org/10.1063/1.1331648

[11] H.M. Rietveld, A Profile Refinement Method for Nuclear and Magnetic structures, J. Appl. Crystallogr. 2 (1969) 65-71
     http://dx.doi.org/10.1107/S0021889869006558

[12] V. Petricek, M. Dusek, L. Palatinus, Jana 2006, The Crystallographic Computing System, Institute of Physics, Prague, Czech Republic, (2006)

[13] Maxim V. Kuznetsov, Ivan P. Parkin, Daren J. Caruanab and Yuri G. Morozova, Combustion synthesis of alkaline-earth substituted lanthanum manganites; $LaMnO_3$, $La_{0.6}Ca_{0.4}MnO_3$ and $La_{0.6}Sr_{0.4}MnO_3$, J. Mater. Chem, 14 (2004) 1377–1382
     http://dx.doi.org/10.1039/b313553p

[14] R.D. Shannon, Revised effective ionic radii and systematic studies of interatomic distances in halides and chalcogenides. Acta Cryst. A32, (1976) 751-767
     http://dx.doi.org/10.1107/S0567739476001551

[15] R. Saravanan, GRAIN software, Private Communication, (2008)

[16] B.D. Culllity, S.R. Stock, Elements of X-ray Diffraction, third ed. Prentice Hall, New Jersy, 2001

[17] A.L.A. da Silva, L. da Conceição, A.M.Rocco, M.M.V.M.Souza, Synthesis of Sr-doped $LaMnO_3$ and $LaCrO_3$ powders by combustion method: structural characterization and thermodynamic evaluation, Cerâmica 58 (2012) 521-528
     http://dx.doi.org/10.1590/S0366-69132012000400018

[18] D.M. Collins, Electron density images from imperfect data by iterative entropy maximization, Nature 49 (1982) 298
     http://dx.doi.org/10.1038/298049a0

[19]   A.D. Ruben, I. Fugio, Superfast program PRIMA for the Maximum Entropy Method, Advanced Materials Laboratory, National Institute for Material Science, Ibaraki, Japan (2004), 3050044

[20]   K. Momma, F. Izumi, VESTA: a three-dimensional visualization system for electronic and structural analysis, J. Applied Crystallogr. 41 (2008) 653-658
        http://dx.doi.org/10.1107/S0021889808012016

[21]   S. Faaland, K.D. Knudsen, M.A. Einarsrud, L. Rormark, R. Hoier, and T. Grande Structure, Stoichiometry, and Phase Purity of Calcium Substituted Lanthanum Manganite Powders, Journal of solid state chemistry 140, (1998) 320-330
        http://dx.doi.org/10.1006/jssc.1998.7894

[22]   J. Tauc, R. Grigorvici, A. Vancu, Optical Properties and Electronic Structure of Amorphous Germanium, Physica Status Solidi 15, 627–637 (1966).
        http://dx.doi.org/10.1002/pssb.19660150224

[23]   H. Kobori, A. Hoshino, A. Yamasaki, A. Sugiura, T. Taniguchi, T Horie, Y. Naitoh, and T. Shimizu, Magneto-resistance enhancement due to self-hole-doping in $LaMnO_3$ produced by low temperature heat treatment, doi:10.1088/1742-6596/400/4/042035

[24]   J.R. Gebhardt, S. Roy, and N. Alia, Colossal magnetoresistance in Ce doped manganese oxides, J. Appl. Phys. 85 (1999) 5390-5392
        http://dx.doi.org/10.1063/1.369987

CHAPTER 14

Synthesis and Charge Density Analysis of BaTiO$_3$

S. Sasikumar, R. Saravanan

Research Centre & PG Department of Physics, The Madura College, Madurai-625011, Tamil Nadu, India

Email: sasikuhan@gmail.com; saragow@gmail.com

Abstract

Perovskite BaTiO$_3$ ceramics were prepared by the solid state reaction technique. Structural and surface morphological properties were explored using the XRD and SEM techniques. The results of structural characterization show the formation of BaTiO$_3$ sample in a single phase without any impurity. It was found to crystallize in a tetragonal symmetry with a P4mm space group. The charge density distribution of BaTiO$_3$ has been constructed using the maximum entropy method. Scanning electron microscope images were used to find the aggregate particle size. The optical absorption analysis was done using a UV-Visible spectrophotometer.

Keywords

Rietveld Refinement, Charge Density, Maximum Entropy Method, Scanning Electron Microscopy

Contents

1.   INTRODUCTION

Barium titanate (BT) is one of the most common ferroelectrics with a wide range of ferroelectric properties, such as spontaneous polarization, high dielectric permittivity, as well as piezo- and pyroelectricity [1, 2]. The ferroelectric properties of BT can be efficiently controlled by doping with different doping elements [3]. It is possible to tailor the parameters like maximum dielectric constant (ε_m), transition temperature (T_c) and $d\varepsilon/dT$ with suitable doping. Reports [e.g. 4] show that the TiO_6 octahedra are disturbed when B-site atoms are replaced with suitable dopant atoms resulting in a broadening of the transition at Tc. Specifically doping with 3d transition elements in BT stabilizes a different structural configuration in the system [5]. Transition metal ion substitution at the Ti site with oxygen vacancy compensation is confirmed, although their behavior is less accurately reproduced.

Barium titanate has important applications in making different types of multilayer capacitors where compositions are doped heterogeneously to introduce compositional gradients that control their temperature dependence [6,7]. The increased pressure for the miniaturization of electronic devices has generated a need for nano-sized ferroelectric powders. The structure and dielectric property of ferroelectric materials are also greatly influenced by their particle size [11–13]. Thus, it is important to examine the formation and dielectric properties of transition metal ion doped nano-sized BT powder.

Among a number of ferroelectrics, barium titanate ($BaTiO_3$) is one of the most basic and widely applied ferroelectric oxide materials with a perovskite-type crystalline structure [14]. With decreasing temperature, it undergoes three successive phase transitions: from the cubic phase to the tetragonal phase (para-electric), then to the orthorhombic phase (ferro-electric) and finally to the rhombohedral phase (ferro-electric) [15]. Due to its high dielectric constant, attractive positive thermal coefficient of resistivity, excellent piezoelectric properties, the application of $BaTiO_3$ in multilayer ceramic capacitors, thermistor and piezoelectric transducers has been extensively investigated [16]. Most related studies of $BaTiO_3$ focus on bulk ceramics [17]. For bulk ceramics, solid state reaction processes, such as microwave sintering and two-step sintering, were adopted [18, 19]. Barium titanate is widely utilized in dynamic random access memory (DRAM) and piezo electric sensors due to its excellent properties, such as high permittivity, outstanding ferro-electric and piezo electric properties, etc. [20-22]. In the present study, the structural and microstructure properties of $BaTiO_3$ are investigated. The reconstruction of the charge density in the $BaTiO_3$ unit cell has been analysed.

2. EXPERIMENTS

2.1 SYNTHESIS

$BaTiO_3$ ceramic was synthesized by the solid state reaction method using high purity analytical grade $BaCO_3$ (99.99%) and TiO_2 (99.99%) as starting materials. The raw materials were weighed with their stoichiometric molar ratio. The weighed powders were milled with an agate ball mill for 3 h. The resulting powders were pelletized. The disk samples were sintered at 1350 °C for 2 h in air atmosphere at the rate of 5 °C/min.

2.2 CHARACTERIZATION

The structure of the sample was analyzed by powder X-ray diffraction (PXRD) using a Bruker AXS D8 Advance (Karlsruhe, Germany). The X-ray diffraction data were measured over the scattering angle range of 10° to 120° at 2θ step of 0.02° with CuKα radiation of wavelength λ=1.54056 Å. Scanning electron microscopy (JEOL Model JSM - 6390LV) was employed to investigate the morphology and particle size in the specimens. The optical absorption spectrum of $BaTiO_3$ was recorded using a UV-Visible spectrometer.

3. RESULTS AND DISCUSSION

3.1 POWDER XRD DATA AND RIETVELD REFINEMENT

The X-ray powder diffraction patterns of the specimens sintered at 1350 °C are shown in figure 1. The XRD Bragg reflection was assigned to that of the tetragonal perovskite structure of $BaTiO_3$. The observed X-ray peaks for the prepared sample were matched with the standard pattern XRD data set from the Joint Committee for Powder Diffraction Standards (JCPDS) reported in the file (JCPDS No: 05-0626).

In the Rietveld refinement [23] analysis, all structural and profile parameters were refined using the software package JANA2006 [24]. The lattice parameter, fractional atomic coordinates, atomic displacement parameters and occupation factors were obtained from whole powder diffraction patterns. $BaTiO_3$ is a FCC cubic structure with space group

P4mm having 4 molecules per unit cell. The fitted profile for XRD data is shown in figure 2. The refined structural and other parameters are tabulated in table 1.

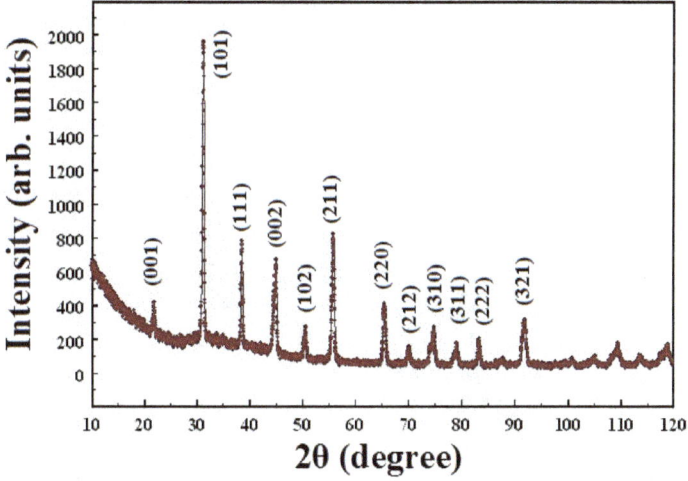

Figure 1. Powder X- ray profile for BaTiO₃.

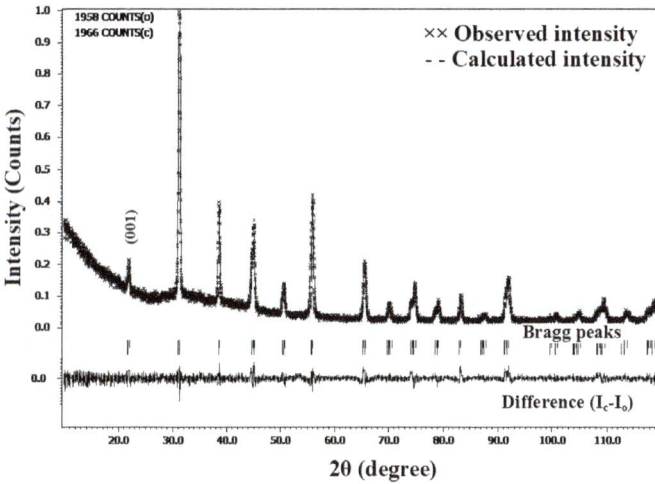

Figure 2. Fitted powder XRD profile for BaTiO₃.

Table 1. Structural Parameters of BaTiO₃.

Parameter	Value
a=b (Å)	3.9924(12)
c (Å)	4.0276(11)
α=β=γ	90°
Density (gm/cc)	6.0299(36)
Volume (Å3)	64.203(39)
F(000)	102
Robs (%)	1.28
Rp (%)	5.63
GOF	1.08

3.2   CHARGE DENSITY USING MAXIMUM ENTROPY METHOD

The refined structure factors extracted from the Rietveld refinement [23] were used to construct the electron density distributions. The electron density of the BaTiO₃ system was visualized by the visualization software VESTA [25]. Figure 3 shows the electron density distribution in the unit cell of BaTiO₃ drawn with the iso-surface level of 0.8. Figure 4 shows the charge density in two dimensional planes drawn from 0 to 2 e/Å³ at an interval of 0.1 e/Å³ along the (100), the (110) and the (200) planes for the prepared sample. The two dimensional charge density drawn along the bonding plane authenticates the increase of charges in the bonding region for the BaTiO₃. To quantify these results, one dimensional electron density line profiles between Ba and O, Ti and O atoms are presented in figure 5. The mid bond electron densities are evaluated and the values along the Ba-O bonding and Ti-O bonding are 0.1439 and 0.4326 e/Å³ respectively. The values of the bond critical points give the nature of bonding between the atoms. The mid bond density values and bond lengths corresponding to Ba-O and Ti-O are given in table 2.

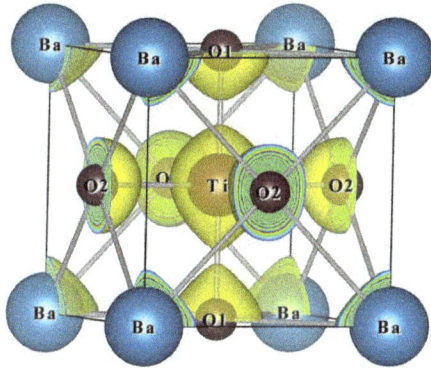

Figure 3. Three dimentional charge density structure of BaTiO₃.

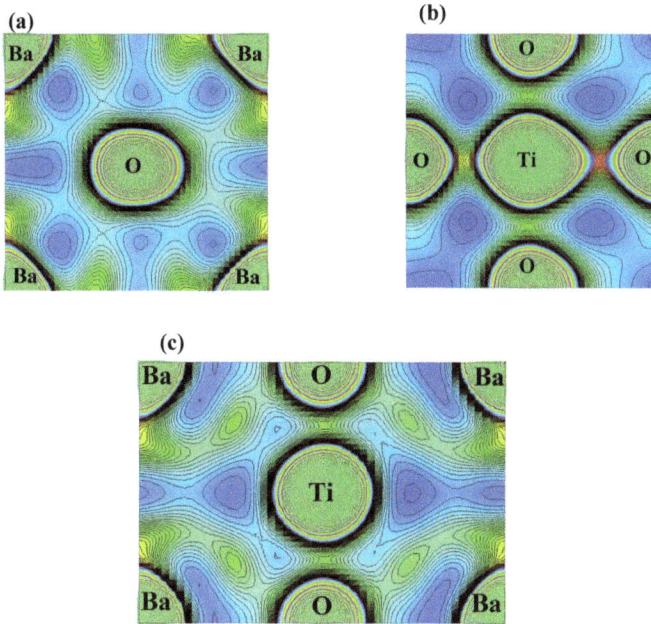

Figure 4. Two dimensional electron density distributions of BaTiO₃ on the a) (100) b) (200) c) (101) plane.

195

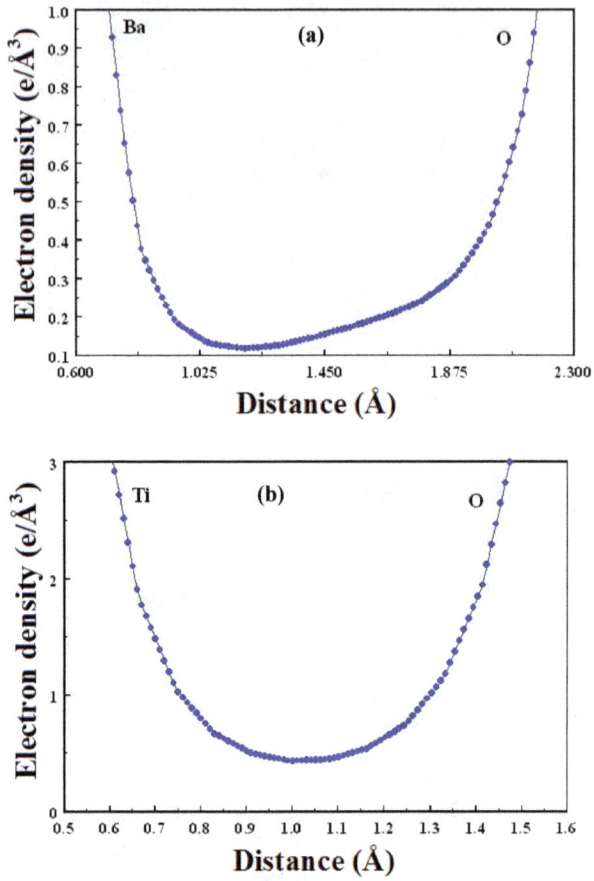

Figure 5 One dimensional electron density profiles for BaTiO₃ a) Ba-O bond b) Ti-O bond.

Table 2 Bond lengths and mid bond electron densities for Ba-O and Ti-O bonds for BaTiO₃.

Ba-O		Ti-O	
Bond length (Å)	Mid bond electron density (e/Å^3)	Bond length (Å)	Mid bond electron density (e/Å^3)
2.8356	0.1439	1.9962	0.4326

3.3   SCANNING ELECTRON MICROSCOPY ANALYSIS

The analysis on the range of the size in the prepared samples was done using a scanning electron microscope. Average particle size of the $BaTiO_3$ is found to be 2.5 μm. Figure 4 shows the micrograph of the prepared barium titanate powders.

Figure 4. SEM micrograph of BaTiO₃.

3.4 UV-VISIBLE ANALYSIS

Absorption spectra were obtained at room temperature in the range of 200 - 2000 nm. In order to study the effect of strain on the energy band gap and absorption peaks, we performed optical absorption spectrum studies. An estimate of the optical band gap is obtained using the below equation for a $BaTiO_3$. According to the equation $(\alpha h\nu)^2 = A (h\nu - E_g)$, [26, 27] and the band gap for $BaTiO_3$ was calculated to be 3.1179 eV and the corresponding graph is shown in figure 5

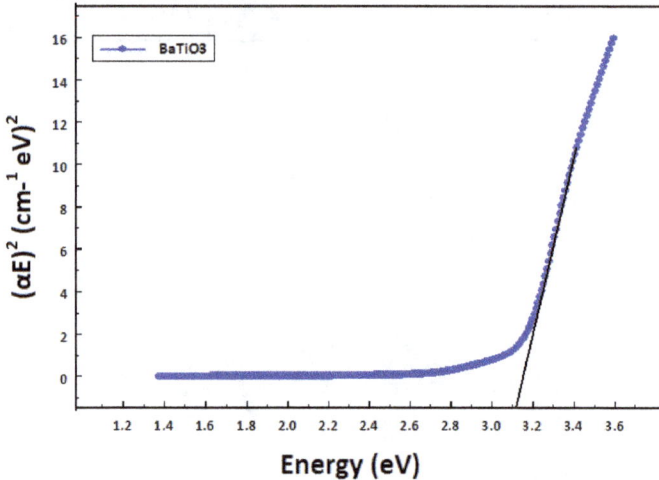

Figure 5. UV-Visible spectra for BaTiO₃.

CONCLUSION

Perovskite $BaTiO_3$ ceramic was synthesized using the solid state reaction method and characterized by PXRD for analyzing its structural information. The electron density distribution in the unit cell is determined for the prepared samples using PXRD through the MEM technique. The morphological studies and average particle size were evaluated by using scanning electron microscopy. The bandgap for the prepared samples were estimated by UV-Visible spectra.

REFERENCES

[1]   B. Jaffe, W. Cook, H. Jaffe, Piezoelectric Ceramics, Academic Press, London, 1971.

[2]   Z. Jing, Z. Yu, C. Ang, Crystalline structure and dielectric properties of Ba(Ti$_{1-y}$Ce$_y$)O$_3$, J. Mater. Sci. 38 (2003) 1057-1061. *http://dx.doi.org/10.1023/A:1022306132064*

[3]   F. Jona, G. Shirane, Ferroelectric Crystal, Dover Publication, New York, 1993.

[4]   D. Hennings, A. Schnell, G. Simon, Diffuse Ferroelectric Phase Transitions in Ba(Ti$_{1-y}$Zr$_y$)O$_3$ Ceramics, J. Am. Ceram. Soc. 65 (1982) 536-544. *http://dx.doi.org/10.1111/j.1151-2916.1982.tb10778.x*

[5]   H.T. Langhammer, T. Muller, K. Felgner, H. Abicht, J. Am. Ceram. Soc. 83 (2000) 605-611. *http://dx.doi.org/10.1111/j.1151-2916.2000.tb01239.x*

[6]   M.E. Lines, A.M. Glass, Principles and Application of Ferroelectric and Related Materials, Oxford University Press, Oxford, 1977.

[7]   A. Rae, M. Chu, V. Ganine, Barium titanate-past, present and future Ceram. Trans. 100 (1999) 1-12.

[8]   K. Albertsen, Reoxidation of Ni-MLCC, J. Eur. Ceram. Soc. 24 (2004) 1883-1887 *http://dx.doi.org/10.1016/S0955-2219(03)00463-1*

[9]   W.H. Tzing, W.H. Tsuan, H.L. Lin, Ceram. Int. 25 (1999) 425. *http://dx.doi.org/10.1016/S0272-8842(98)00058-3*

[10]   W.H. Tzing, W.H. Tsuan, Ceram. Int. 25 (1999) 69. *http://dx.doi.org/10.1016/S0272-8842(98)00003-0*

[11]   G. Arlt, D. Hennings, G. de With, Dielectric properties of fine-grained barium titanate ceramics, J. Appl. Phys. 58 (1985) 1619. *http://dx.doi.org/10.1063/1.336051*

[12]   K. Uchino, E. Sadanaga, T. Hirose, Dependence of the crystal structure on particle size in barium titanate, J. Am. Ceram. Soc. 72 (1989) 1555-1558. *http://dx.doi.org/10.1111/j.1151-2916.1989.tb07706.x*

[13]   T.T. Fang, H.L. Hseih, F.S. Shiau, Effect of Pore Morphology and Grain Size on the Dielectric Properties and Tetragonal-Cubic Phase Transition of High-Purity Barium Titanate, J. Am. Ceram. Soc. 76 (1993) 1205-1211. *http://dx.doi.org/10.1111/j.1151-2916.1993.tb03742.x*

[14]   A. Kirchner, M. Arin, P. Lommens, X. Granados, S. Ricart, B. Holzapfel, I. Van Driessche, Chemical solution deposition of multiferroic La$_{0.7}$Sr$_{0.3}$MnO$_3$, BaTiO$_3$ thin films prepared by ink plotting, J. Alloys Compd. 516 (2012) 16–19. *http://dx.doi.org/10.1016/j.jallcom.2011.11.025*

[15] P. Zheng, J.L. Zhang, Y.Q. Tan, C.L. Wang, Grain-size effects on dielectric and piezoelectric properties of poled $BaTiO_3$ ceramics Acta Mater. 60 (2012) 5022–5030.
http://dx.doi.org/10.1016/j.actamat.2012.06.015

[16] P. Ren, H. Fan, X. Wang, K. Liu, A novel approach to prepare tetragonal $BaTiO_3$ nanopowders, Mater Lett. 65 (2011) 212–214.
http://dx.doi.org/10.1016/j.matlet.2010.10.015

[17] S. Chao, F. Dogan, $BaTiO_3$-$SrTiO_3$ Layered Dielectrics for Energy Storage Mater Lett. 65 (2011) 978–981.

[18] N. Raengthon, T. Sebastian, D. Cumming, Ian. M. Reaney, D.P. Cann, $BaTiO_3$–$Bi(Zn_{1/2}Ti_{1/2})$ O_3–$BiScO_3$ Ceramics for High-Temperature Capacitor Applications J. Am. Ceram. Soc. 95 (2012) 3554–3561.
http://dx.doi.org/10.1111/j.1551-2916.2012.05340.x

[19] Y. Wang, X. Chen, H. Zhou, L. Fang, L. Liu, H. Zhang, J. Alloys Compd. 551 (2013) 365–369.
http://dx.doi.org/10.1016/j.jallcom.2012.09.127

[20] C. Chen, Y. Wei, X. Jiao, D. Chen, Hydrothermal synthesis of $BaTiO_3$: crystal phase and the Ba^{2+} ions leaching behavior in aqueous medium, Materials Chemistry and Physics 110 (2008) 186–191.
http://dx.doi.org/10.1016/j.matchemphys.2008.01.031

[21] H. Zhang, X. Wang, Z. Tian, C. Zhong, Y. Zhang, C. Sun, L. Li, Journal of the American Ceramic Society 94 (2011) 3220–3222.
http://dx.doi.org/10.1111/j.1551-2916.2011.04805.x

[22] Y. Wang, G. Xu, L. Yang, Z. Ren, X. Wei, W. Weng, P. Du, G. Shen, G. Han, Materials Letters 63 (2009) 239–241.
http://dx.doi.org/10.1016/j.matlet.2008.09.050

[23] H.M. Rietveld, A Profile Refinement Method for Nuclear and Magnetic, J. Appl. Crystallogr. 2 (1969) 65.
http://dx.doi.org/10.1107/S0021889869006558

[24] V. Petricek, M. Dusˇek, L. Palatinus, nin: JANA2000, The Crystallographic Computing System, Institute of Physics, Academy of Sciences of the Czech Republic, Praha, (2000).
http://dx.doi.org/10.1038/298049a0

[25] D.M. Collins, Electron density images from imperfect data by iterative entropy maximization Nature, 298 (1982) 49-51.

[26] J. Tauc, R. Grigorvici, and Y. Yanca, Optical Properties and Electronic Structure of Amorphous Germanium, Phys. Status Solidi 15 (1966) 627 -637
http://dx.doi.org/10.1002/pssb.19660150224

[27] J. Pancove, Optical Processes in Semiconductors (Englewood Cliffs, NJ: Prentice-Hall, 1979).

CHAPTER 15

Synthesis and Structural Characterizations of $Na_{1-x}K_xNb_{0.95}Sb_{0.05}O_3$

S. Sasikumar, R. Saravanan

Research Centre and Post Graduate Department of Physics, The Madura College, Madurai-625 011, Tamil Nadu, India

Email: saragow@gmail.com; sasikuhan@gmail.com

Abstract

Solid solutions of lead free ceramics of $Na_{1-x}K_xNb_{0.95}Sb_{0.05}O_3$ (x=0.01, 0.03 and 0.05) were prepared by the conventional solid state reaction method. The formation of perovskite structure of the prepared ceramics was confirmed by means of room temperature powder X-ray diffraction. X-ray powder diffraction analyses revealed the persistence of an orthorhombic phase. The crystal structures were refined using profile refinement. Aggregated average particle sizes were evaluated using scanning electron microscopy. Elemental compositional analysis was carried out using energy dispersive spectrum.

Keywords

Crystal Structure, X-Ray Diffraction, Rietveld Analysis, Scanning Electron Microscopy.

Contents

1. INTRODUCTION

Lead-based Pb(Zr Ti)O$_3$ piezoelectric ceramics are widely used in piezoelectric devices such as actuators and sensors [1,2], surface acoustic wave filters (SAW) [3], transducers [4], buzzers [5], and piezoelectric transformers [6,7] due to their excellent electrical properties and temperature stability. However, lead-based piezoelectric ceramics cause environmental pollution and health problems because of the high toxicity of PbO evaporation. A lot of countries have recently limited the use of lead-based products. A replacement for lead based piezoelectric ceramics is thus required. Many lead-free piezoelectric ceramic systems have been studied, such as Bi compounds [8, 9], (Na, K)NbO$_3$ [10,11], and BaTiO$_3$ [12,13]. Among lead-free piezoelectric ceramics, (Na, K)NbO$_3$-based ceramic systems have attracted a lot of attention as replacements for lead-based ceramics due to their excellent piezoelectric properties, high Curie temperature, and relatively low harm to the environment. It is well known that pure NKN ceramics are difficult to synthesize using the conventional solid state reaction method through an ordinary sintering process and that they easily decompose when exposed to air and moisture due to the evaporation of alkaline elements under high sintering temperature. When the sintering temperature is above 1000 °C, the evaporation of K$_2$O and Na$_2$O degrades the resistivity and the piezoelectric properties. However, it is very difficult to control the evaporation of Na$_2$O and K$_2$O by muffling. Therefore, new methods are needed to maintain the electrical and piezoelectric properties of NKN-based samples, such as adjusting the K/Na ratio or A/B site ratio for the optimal proportion [14–17] or lowering the sintering temperature to under 1000 °C by doping the samples with extrinsic materials [18]. In our work, we report on structural and morphological properties of Na$_{1-x}$K$_x$Nb$_{0.95}$Sb$_{0.05}$O$_3$ (x=0.01, 0.03 and 0.05).

2. SAMPLE PREPARATION

The lead free ceramics Na$_{1-x}$K$_x$Nb$_{0.95}$Sb$_{0.05}$O$_3$ (x=0.01, 0.03 and 0.05) have been prepared by the solid state reaction method. The raw compounds of Na$_2$CO$_3$ (purity 99.99%), K$_2$CO$_3$ (purity 99.99%), Nb$_2$O$_5$ (purity 99.99%), Sb$_2$O$_3$ (99.99%) were weighted according to their stoichiometric ratio and ground using a ball mill for 3 hrs. The mixed powders were calcined at 900 °C for 4 hrs in air. The raw compounds were kept in alumina crucible with closed lid in tubular furnace. The calcined powders were sintered at 1110 °C for 2 hrs in a programmable furnace. The synthesized powders were structurally characterized by powder X-ray diffraction using CuKα (λ=1.54056Å) radiation. The X-ray diffraction data were measured over the scattering angle range $10°$ to $120°$ at 2θ step of $0.02°$ using CuKα radiations at room temperature (Bruker D8, Karlsruhe, Germany). The surface morphology of the samples was recorded using scanning electron microscopy (SEM). The energy dispersive spectroscopy (EDS) analysis of the composites is also done to investigate the composition.

3. RESULTS AND DISCUSSION

3.1 X-RAY DIFFRACTION AND PROFILE REFINEMENT ANALYSIS

Figure 1 shows the X-ray diffraction (XRD) patterns of the $Na_{1-x}K_xNb_{0.95}Sb_{0.05}O_3$ (x=0.01, 0.03 and 0.05). The XRD patterns of the $Na_{1-x}K_xNb_{0.95}Sb_{0.05}O_3$ are indexed using the standard JCPDS data on $NaNbO_3$ (JCPDS No # 05-0626) and reveal an orthorhombic structure. Any extra or impurity peak is not detected within the detection limit of the XRD. Our results suggest that K^+ and Sb^{5+} have diffused into the $NaNbO_3$ lattices, with K^+ entering the Na^+ sites and Sb^{5+} occupying the Nb^{5+} sites, to form a homogeneous solid solution. Figure 1(b) shows the enlarged (200) peak, which shows that the peak shift towards the lower diffraction angles side of 2θ diffracting angle. It can be seen that the unit cell volume increased with the increase of K^+ content on $NaNbO_3$. It is expected that K^+ (1.33Å) should substitute for Na^+ (1.02Å) at the A site [19]. The substitution on the A-site may be leading to the expansion of the crystal unit cell. The increasing trend in the cell volume with the increase in K^+ concentration indicates the replacement on $NaNbO_3$ by potassium. The variations in the lattice parameters and cell volume have also been studied for different doping concentrations and are presented in table 1.

The lattice parameters of the composites are calculated using the JANA 2006 software [20], based on the Rietveld refinement method [21]. This technique requires an approximate structural model of the real structure. For the refinement, various structural parameters (such as background, scale factors, profile half-width parameters, atomic coordinates, lattice parameters) are varied during the successive refinement cycles, in order to obtain a good match between the observed and the calculated diffraction pattern. The reliability of the refinement is determined by a parameter known as "goodness of fit". Figures 2 (a)-(f) confirm a good agreement between observed XRD patterns and theoretical fits which indicate the success of the Rietveld refinement [21] method. The values of the lattice parameters are listed in table 1. It can be noticed from table 1 that the values of the lattice parameters obtained for the $Na_{1-x}K_xNb_{0.95}Sb_{0.05}O_3$ samples show slight variation when compared with other compositions.

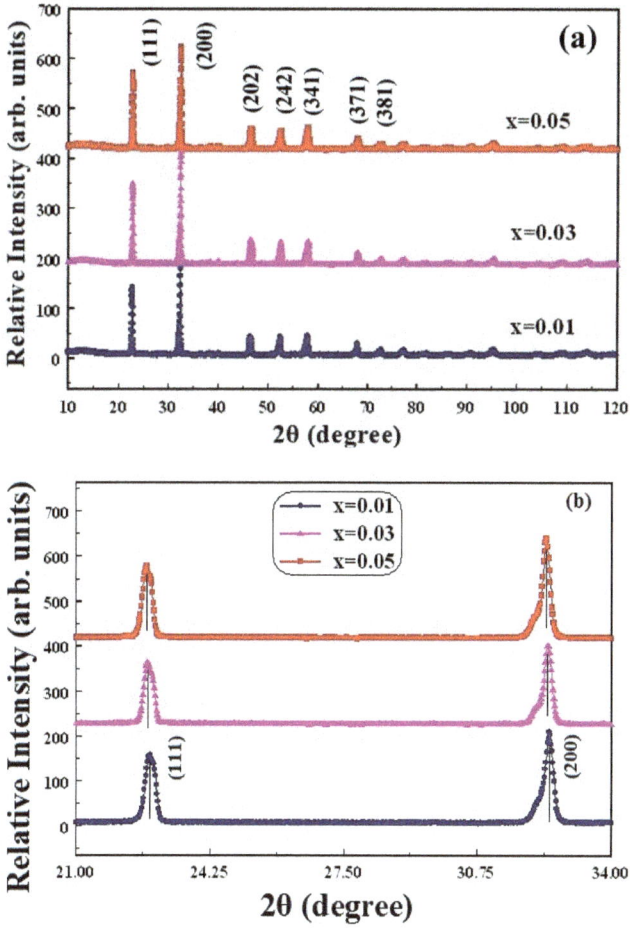

Figure 1 (a) X-diffraction patterns for $Na_{1-x}K_xNb_{0.95}Sb_{0.05}O_3$ (b) enlarged (2 0 0) peak.

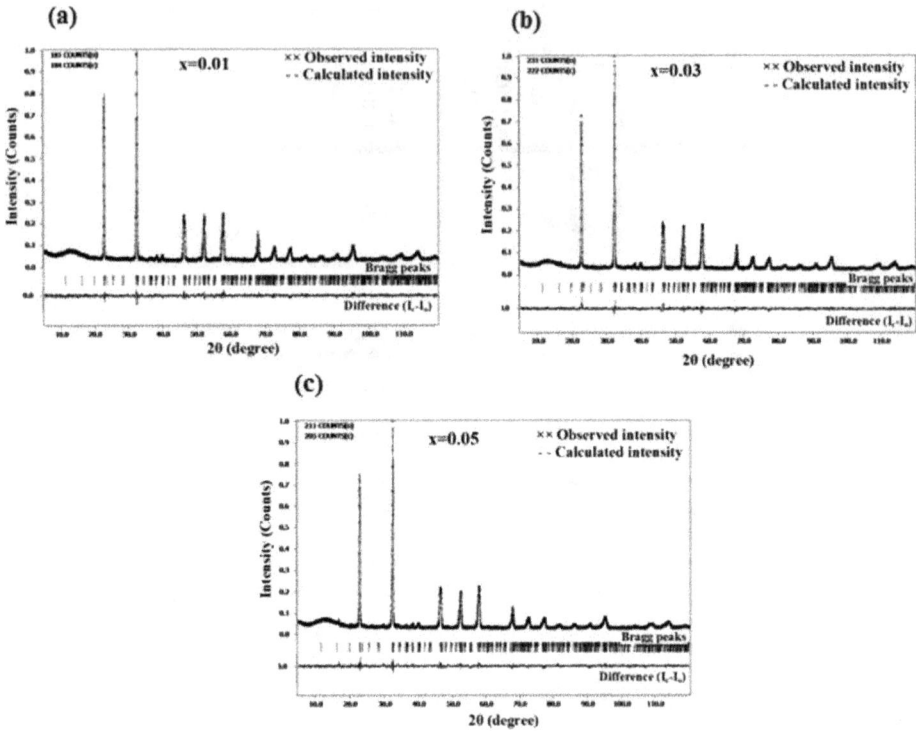

Figure 2 Rietveld refinement profile for $Na_{1-x}K_xNb_{0.95}Sb_{0.05}O_3$ a) x=0.01, b) x=0.03, c) x=0.05.

Table 1 Refined structural parameters from Rietveld refinement of $Na_{1-x}K_xNb_{0.95}Sb_{0.05}O_3$ (x=0.01, 0.02, 0.03).

Parameters	x=0.01	x=0.03	x=0.05
a (Å)	5.5055(4)	5.5000(4)	5.5152(7)
b (Å)	5.5618(3)	5.5522(1)	5.5681(8)
c (Å)	15.5484(8)	15.5313(3)	15.5705(7)
α=β=γ (°)	90	90	90
Volume (Å³)	476.11 (6)	474.29(4)	478.16(4)
Density (gm/cc)	4.61 (1)	4.64(2)	4.61(5)
Rp (%)	5.52	5.59	5.44
Robs (%)	2.75	3.15	2.26
GOF	0.21	0.23	0.22
F(000)	613	614	615

3.2 MICROSTRUCTURAL ANALYSIS

The surface morphology of the grown samples was recorded using scanning electron microscopy (SEM). The SEM images of the sintered powders of all the samples are presented in figure 3 (a-c). From SEM micrographs, it is clearly observed that the particle sizes are decreasing with addition of dopants. The energy dispersive spectroscopy (EDS) analysis of the composites is also done to investigate the composition. The percentages of elements present in compositions are listed in table 2. The EDS analysis supports the XRD results. EDS spectra of the NKNS ceramics were shown in figure 3 (d-f). It is noticed that the elements present in the solid solution matched well with their nominal stoichiometry. The results indicate that the constituent ions are present in the respective samples in expected proportions.

Figure 3 SEM micrographs of the prepared $Na_{1-x}K_xNb_{0.95}Sb_{0.05}O_3$, a) x=0.01 b) x=0.03 c) x=0.05 ceramics with their elemental composition found using EDAX, (d) x=0.01 (e) x=0.03 (f) x=0.05.

Table 2 Elemental composition of the $Na_{1-x}K_xNb_{0.95}Sb_{0.05}O_3$ ceramics determined by EDS analysis.

Compositions	Atomic percent (%)					Weight percent (%)				
	Na	K	Nb	Sb	O	Na	K	Nb	Sb	O
x=0.01	41.63	0.35	35.58	1.45	20.99	19.98	0.28	69.02	3.70	7.01
x=0.03	41.62	0.31	35.61	1.49	20.97	19.96	0.26	69.01	3.78	7.00
x=0.05	39.39	1.11	36.67	1.69	20.55	18.74	0.89	69.47	4.19	6.70

CONCLUSION

The solid solution of lead free piezo electric ceramics $Na_{1-x}K_xNb_{0.95}Sb_{0.05}O_3$ (x=0.01, 0.03 and 0.05) were prepared by the solid state reaction method. The structural and morphological properties of $Na_{1-x}K_xNb_{0.95}Sb_{0.05}O_3$ ceramic samples have been studied. X-ray powder diffraction confirms the formation of orthorhombic single phase crystalline structure without any additional phase content. All XRD patterns were analyzed by using a profile refinement technique which revealed the orthorhombic structure. EDS analysis shows that the quantities of the elements present in the sample increase with composition x.

REFERENCES

[1]   T.-Y. Chen, S.-Y. Chu, C.-K. Cheng, Doping effects on the piezoelectric properties of low-temperature sintered PbTiO3-based ceramics for SAW applications, Integrated Ferroelectrics 58 (2003) 1315–1324.
http://dx.doi.org/10.1080/714040790

[2]   S.-Y. Chu, T.-Y. Chen, W. Water, The investigation of preferred orientation growth of ZnO films on the PbTiO3-based ceramics and its application for SAW devices, Journal of Crystal Growth 257 (2003) 280–285.
http://dx.doi.org/10.1016/S0022-0248(03)01452-0

[3]   S.-Y. Chu, T.-Y. Chen, I.T. Tsai, Effects of poling field on the piezoelectric and dielectric properties of Nb additive PZT-based ceramics and their applications on SAW devices, Materials Letters 58 (2004) 752–756.
http://dx.doi.org/10.1016/j.matlet.2003.07.004

[4] B. Jadidian, N. Hagh, A. Winder, A. Safari, 25 MHz ultrasonic transducers with lead-free piezoceramic, 1-3 PZT fiber-epoxy composite, and PVDF polymer active elements, IEEE Transactions on Ultrasonics Ferroelectrics and Frequency Control 56 (2009) 368–378.
http://dx.doi.org/10.1109/TUFFC.2009.1046

[5] K.-T. Chang, H.-C. Chiang, C.-W. Lee, Design and implementation of a piezoelectric clutch mechanism using piezoelectric buzzers, Sensors and Actuators A: Physical 141 (2008) 515–522.
http://dx.doi.org/10.1016/j.sna.2007.10.018

[6] Z. Yang, X. Chao, R. Zhang, Y. Chang, Y. Chen, Fabrication and electrical characteristics of piezoelectric PMN-PZN-PZT ceramic transformers, Materials Science and Engineering B: Solid-State Materials for Advanced Technology 138 (2007) 277–283.
http://dx.doi.org/10.1016/j.mseb.2007.01.031

[7] C.-W. Ahn, H.-C. Song, S. Nahm, S. Priya, S.-H. Park, K. Uchino, H.-G. Lee, H.-J. Lee, Effect of ZnO and CuO on the sintering temperature and piezoelectric properties of a hard piezoelectric ceramic, Journal of the American Ceramic Society 89 (2006) 921–925.
http://dx.doi.org/10.1111/j.1551-2916.2005.00823.x

[8] T. Takenaka, T. Gotoh, S. Mutoh, T. Sasaki, New series of bismuth layer structured ferroelectrics, Japanese Journal of Applied Physics Part 1: Regular Papers Short Notes and Review Papers 34 (1995) 5384–5388.
http://dx.doi.org/10.1143/JJAP.34.5384

[9] M. Hirose, T. Suzuki, H. Oka, K. Itakura, Y. Miyauchi, T. Tsukada, Piezoelectric properties of $SrBi_4Ti_4O_{15}$-based ceramics, Japanese Journal of Applied Physics Part 1: Regular Papers Short Notes and Review Papers 38 (1999) 5561–5563.
http://dx.doi.org/10.1143/JJAP.38.5561

[10] Y. Saito, H. Takao, T. Tani, T. Nonoyama, K. Takatori, T. Homma, T. Nagaya, M. Nakamura, Lead-free piezoceramics, Nature 432 (2004) 84– 87.
http://dx.doi.org/10.1038/nature03028

[11] J. Acker, H. Kungl, M.J. Hoffmann, Influence of alkaline and niobium excess on sintering and microstructure of sodium–potassium niobate $(K_{0.5}Na_{0.5})NbO_3$, Journal of the American Ceramic Society 93 (2010) 1270–1281.
http://dx.doi.org/10.1111/j.1551-2916.2010.03578.x

[12] O.P. Thakur, C. Prakash, A.R. James, Enhanced dielectric properties in modified barium titanate ceramics through improved processing, Journal of Alloys and Compounds 470 (2009) 548–551.
http://dx.doi.org/10.1016/j.jallcom.2008.03.018

[13] C.-W. Ahn, D. Maurya, C.-S. Park, S. Nahm, S. Priya, A generalized rule for large piezoelectric response in perovskite oxide ceramics and its application for design of lead-free compositions, Journal of Applied Physics 105 (2009).
http://dx.doi.org/10.1063/1.3142442

[14] S.-Y. Chu, W. Water, Y.-D. Juang, J.-T. Liaw, Properties of (Na, K)NbO$_3$ and (Li, Na, K)NbO$_3$ ceramic mixed systems, Ferroelectrics 287 (2003) 23–33.
http://dx.doi.org/10.1080/00150190390200767

[15] Q. Zhang, B. Zhang, H. Li, P. Shang, Effects of Na/K ratio on the phase structure and electrical properties of Na$_x$K$_{1-x}$NbO$_3$ lead-free piezoelectric ceramics, Rare Metals 29 (2010) 220–225.
http://dx.doi.org/10.1007/s12598-010-0038-y

[16] W. Jiagang, X. Dingquan, W. Yuanyu, W. Lang, J. Yihang, Z. Jianguo, K/ Na ratio dependence of the electrical properties of [(K$_x$Na$_{1-x}$)0.95-)0.95Li$_{0.05}$](Nb$_{0.95}$Ta$_{0.05}$)O$_3$ lead-free ceramics, Journal of the American Ceramic Society 91 (2008) 2385–2387.
http://dx.doi.org/10.1111/j.1551-2916.2008.02415.x

[17] M. Matsubara, T. Yamaguchi, W. Sakamoto, K. Kikuta, T. Yogo, S.-I. Hirano, Processing and piezoelectric properties of lead-free (K,Na) (Nb,Ta) O$_3$ ceramics, Journal of the American Ceramic Society 88 (2005) 1190–1196
http://dx.doi.org/10.1111/j.1551-2916.2005.00229.x

[18] M.-R. Yang, C.-S. Hong, C.-C. Tsai, S.-Y. Chu, Effect of sintering temperature on the piezoelectric and ferroelectric characteristics of CuO doped 0.95(Na$_{0.5}$K$_{0.5}$)NbO$_3$–0.05LiTaO$_3$ ceramics, Journal of Alloys and Compounds 488 (2009) 169–173.
http://dx.doi.org/10.1016/j.jallcom.2009.07.174

[19] R. D. Shannon, Revised effective ionic radii and systematic studies of interatomic distances in halides and chalcogenides, Acta Cryst. A32 (1976) 751-767.
http://dx.doi.org/10.1107/S0567739476001551

[20] V. Petricek, M. Dusek and L. Palatinus, (2006) Jana, The crystallographic computing system (Institute of Physics), Praha, Czech Republic.

[21]   H.M. Rietveld, A Profile Refinement Method for Nuclear and Magnetic, J. Appl. Crystallogr. 2 (1969) 65.
http://dx.doi.org/10.1107/S0021889869006558

Keywords

About the Editor

Dr Ramachandran Saravanan, has been associated with the Department of Physics, The Madura College, affiliated with the Madurai Kamaraj University, Madurai, Tamil Nadu, India from the year 2000. He is the head of the Research Centre and PG department of Physics. He worked as a research associate during 1998 at the Institute of Materials Research, Tohoku University, Sendai, Japan and then as a visiting researcher at Centre for Interdisciplinary Research, Tohoku University, Sendai, Japan up to 2000.

Earlier, he was awarded the Senior Research Fellowship by CSIR, New Delhi, India, during Mar. 1991 - Feb.1993; awarded Research Associateship by CSIR, New Delhi, during 1994 – 1997. Then, he was awarded a Research Associateship again by CSIR, New Delhi, during 1997- 1998. Later he was awarded the Matsumae International Foundation Fellowship in1998 (Japan) for doing research at a Japanese Research Institute (not availed by him due to the simultaneous occurrence of other Japanese employment).

He has guided six Ph.D. scholars as of 2016, and about ten researchers are working under his guidance on various research topics in materials science, crystallography and condensed matter physics. He has published around 100 research articles in reputed Journals, mostly International, apart from around 45 presentations in conferences, seminars and symposia. He has also guided around 50 M.Phil. scholars and an equal number of PG students for their projects. He has attracted government funding in India, in the form of Research Projects. He has completed two CSIR (Council of Scientific and Industrial Research, Govt. of India), one UGC (University Grants Commission, India) and one DRDO (Defense Research and Development Organization, India) research projects successfully and is proposing various projects to Government funding agencies like CSIR, UGC and DST.

He has written 3 books in the form of research monographs with details as follows; "Experimental Charge Density - Semiconductors, oxides and fluorides" (ISBN-13: 978-3-8383-8816-8; ISBN-10:3-8383-8816-X), "Experimental Charge Density - Dilute Magnetic Semiconducting (DMS) materials" (ISBN-13: 978-3-8383-9666-8; ISBN-10: 3-8383-9666-9) and "Metal and Alloy Bonding - An Experimental Analysis" (ISBN -13: 978-1-4471-2203-6). He has committed to write several books in the near future.

His expertise includes various experimental activities in crystal growth, materials science, crystallographic, condensed matter physics techniques and tools as in slow evaporation, gel, high temperature melt growth, Bridgman methods, CZ Growth, high vacuum sealing etc. He and his group are familiar with various equipment such as: different types of cameras; Laue, oscillation, powder, precession cameras; Manual 4-

circle X-ray diffractometer, Rigaku 4-circle automatic single crystal diffractometer, AFC-5R and AFC-7R automatic single crystal diffractometers, CAD-4 automatic single crystal diffractometer, crystal pulling instruments, and other crystallographic, material science related instruments. He and his group have sound computational capabilities on different types of computers such as: IBM – PC, Cyber180/830A – Mainframe, SX-4 Supercomputing system – Mainframe. He is familiar with various kind of software related to crystallography and materials science. He has written many computer software programs himself as well. Around twenty of his programs (both DOS and GUI versions) have been included in the SINCRIS software database of the International Union of Crystallography.